Here's what recruits are saying about this book!

The information in this book is not only well researched but, also very helpful for any firefighter recruit. I can personally vouch for the interviewing section in the book as it has helped me to get re~~~~~~ ~~ ~~~ Sarnie Fire Deparment. This book is definitely v

I got recruited from Mississauga Fire Departmer
"I'm on cloud nine!" I wanted to drop you a line a~~~ ~~y thanks for all your help". You and your website have been a big help in achieving my dream. Thanks again!
                                                                    ~ Dan Burton

It's so great to have another resource to double check things besides 'word of mouth'. While trying to update my resume, I really appreciated all the help from the résumé review, templates and evaluator. All these tools will benefit me in the recruiting process!   *~ Jessica Thomson*

This guide will help you organize everything you need to apply for a firefighter position. Before I got this book, I had no clue about what to put in my application package and what order it should go in and my resume was horrible. However, the book and the website www.becomingafirefighter.com changed all that for me by reviewing my resume, cover letter and references letter making it 10 times better than before I got the membership…This isn't even the half of it. Think about the amount of money you have spent or will spend on your firefighting training, application fees, fitness tests and aptitude tests. Do yourself a favor and invest in this book. It's worth it.   *~ Scott Tadema*

THE COMPLETE GUIDE
TO BECOMING
A FIREFIGHTER

THE COMPLETE GUIDE TO BECOMING A FIREFIGHTER

by

Kory Pearn

From the Front Line of Firefighting

White Knight Books
Toronto, Canada

Published in 2006 by White Knight Books,
a division of Bill Belfontaine Ltd.
Suite 103, One Benvenuto Place
Toronto Ontario Canada M4V 2L1
T. 416-925-6458 F, 416-925-4165
E-mail whitekn@istar.ca
Web: www.whiteknightbooks.ca

Ordering information – Canada
White Knight Book Distribution Services Ltd.
c/o Georgetown Terminal Warehouses
34 Armstrong Avenue
Georgetown ON, L7G 4R9
T: 866-485-5556 F: 866-485-6665
E-mail: orders@gtwcanada.com

Ordering Information – USA
Hushion House Publishing
Services c/o APG Books
1501 Country Hospital Road
Nashville TN USA 37218
T: 888-275-2606
F: 800-510-3650

Library and Archives Canada Cataloguing in Publication
Pearn, Kory
The complete guide to becoming a firefighter : the ultimate
recruitment guide for Canadian firefighters / Kory Pearn.
Includes bibliographical references.
ISBN 0-9780570-1-5 (pbk.)
1. Fire fighters--Vocational guidance--Canada. I. Title.
TH9119.P43 2006   363.37'802371   C2006-903912-7

Cover and text Design:
Karen Thomas, Intuitive Design International Ltd., Omemee, ON

Cover photo: Firstlight.ca

Editing: Bill Belfontaine

Printed and Bound in Canada

DEDICATION

This book is dedicated to those individuals who want to pursue a career in fire fighting; dedicating their valuable time and effort to become the best fire fighter they can be.

Lives, property and dreams remain alive and well because of the heroes they will become.

ACKNOWLEDGEMENTS

When I wrote the pilot for this book two years ago I never imagined having it published. This book would not have been possible without the love and support that my wife Lisa has unflagginly given throughout this project. Not only are you my cheering section but, your are my inspiration.

To all my family you have always been there—Thanks.

Thank you to the St. Thomas Fire Department for allowing me the freedom to involve the fire department as well as memebers of the fire department in this project. As to those members—Ian Thomas, Scott Brett, Chris Gielen and Daryl Smith—thanks for your efforts and encouragement.

Special thanks to Jeanette Fletcher and Michael Stoparczyk from Studio 19 Photography and Digital Imaging, for capturing the images throughout the book. Unlimited talent; as always, refreshing to work with.

I would also like to thank Bill Belfontaine and White Knight Books for coming into my life and helping to make my publishing dreams a reality.

TABLE OF CONTENTS

ABOUT THE BOOK

Becoming a firefighter is a dream career for many men and women. This book provides an innovative and straightforward guide for converting your dream into reality. I have spent three years carefully writing this book after being recruited by St. Thomas' Fire Department. Not only have I researched this material—I have lived it!

This book does not retrace the steps I took to become a firefighter, but instead acts as a resource tool that guides the reader in becoming a firefighter. Every step to becoming a firefighter are included allowing the reader to start reading at any stage in their journey. The book also features many interactive segments which not only provides information to the reader, but also enables the reader to utilize this information immediately.

Ultimately *The Complete Guide to Becoming a Firefighter* is an invaluable asset to anyone who wants to become a firefighter or is considering such a profession.

HOW TO GET THE MOST
OUT OF THIS BOOK

Just to assure you there is no misunderstanding; I will set the record straight. This book is designed to be a resource tool full of ideas, examples, tips and suggestions to inspire you and to provide vital information to allow you to be in control of your own recruitment process.

How you can get the most out of the time you spend:

- Adopt the ideas and make them your own. Every firefighter recruit is unique in the sense that everyone has different qualities to offer fire departments. Choose the ideas that pertain to you and adapt them to suit your profile.
- As you read this book you will find many new ideas, make sure you do not to get overwhelmed and try to accomplish everything at once. Maintain your focus, once you have figured out a plan; stick to it. Planning is the first and most critical step in any recruitment process.
- Learn, learn, and learn more. That means, constantly resorting back to sections of this book and use it as a reference. Before each stage of your recruitment process, it is important to be refreshed with the tips and ideas you'll find in this book.
- Most important, be sure to continually evaluate your recruitment process and the direction in which it is headed. Always ask yourself if there is something more you could be doing. Are you on the right path? What are other recruits doing? Have you been making progress?

The best way to get the most out of this book is to make notes right on the pages and/or highlight the areas that interest you. Carry this book around so you can refer to it at any time.

Don't make the mistake of reading "The Guide" only once then leaving it aside. Almost half you learned will be forgotten in the first few weeks and you will miss vital information you didn't pick up the first time.

Need additional information on cover letters, résumés, recruitment trends, and recruiting fire departments? Visit Canada's leading interactive recruitment resource: *www.becomingafirefighter.com*

Once you have read this book and understand the fundamentals of becoming a firefighter, this website will become an invaluable tool to every serious firefighter recruit. Apply the knowledge you have learned throughout the book to your recruitment process. Let our website: **www.becomingafirefighter.com**, show you the best study route to follow.

Kory Pearn, author of *"The Complete Guide to Becoming a Firefighter"* manages this website personally. "Members to my website, becomingafirefighter.com receive vital up-to-date information as it happens as well as the ability to contact me any time with questions or concerns regarding your recruitment."

Become a member of:
www.becomingafirefighter.com

Firefighter's Prayer

When I am called to duty, God
Where ever flames may rage
Give me the strength to save some life
Whatever be its age
Help me embrace a little child
Before it's too late
Or save an older person from
The horror of that fate
Enable me to be alert and
Hear the weakest shout
And quickly and efficiently
To put the fire out
I want to fill my calling and
To give the best in me
To guard my every neighbor
And protect their property
And if according to your will
I have to lose my life
Please bless with your protecting hand
My children and my wife.

(Author Unknown)

Part I

Getting Started

CHAPTER ONE

JOB DESCRIPTION OF TODAY'S FIREFIGHTER

One of the most common phrases describing firefighters is *'Always running into a burning building when others are running out!'* This slogan can be found on t-shirts as well as in movies. Even though this saying is over-used doesn't change the fact that it is true. Firefighters are constantly running into structures when all other personal are running out. For firefighters to do their jobs properly it almost always involves risks.

The constant physical demands of the job are infinite; as is the ability to recall years of education and training. Today's firefighter is trained to handle every type of emergency situations such as vehicle accidents, train derailment, hazardous material incidents and as always fires. However, only 20% of the calls firefighters respond to are fires, the rest are medical-related emergencies: heart attacks, strokes, VSA (vital signs absent), seizures, severe bleeding, delay of EMS, home accidents, drowning, and any other emergency that I've seen or you can think of. We must be highly diversified in order to handle every type of emergency and non-emergency situations with little or no warning or preparation; firefighters definitely define the word hero.

Firefighters must adapt to any situation that arises. Firefighters are the end of the road. If firefighters can't handle a situation, then the situation can't be handled. They can't turn around and call someone else. It's critical that firefighters know their jobs and execute them without failure.

Our day is filled with situations that also include rescues and medical emergencies as well as maintenance of tools and equipment, pre-planning, community education, building inspections, personal development/training, station and apparatus maintenance and volunteering in the community. Firefighters must be physically fit to meet the demand and strain that is applied to the body under high heat conditions and all other dangerous environments.

Firefighters have many roles and each one is equally important. Throughout your recruitment keep an open mind about how diverse fire departments must be.

Remember the whole picture especially when you're selecting your courses and preparing your résumé.

Success seems to be largely a matter of hanging on
after others have let go.
— William Feather

ROLES AND RESPONSIBILITIES OF A FIREFIGHTER

A firefighter: Extinguishes fires, protects life and property, and operates and maintains equipment by performing the following duties:

- Create and/or maintain access into structures.
- Ventilate structures using natural, mechanical, vertical or horizontal ventilation.
- Respond to medical emergencies—chest pains, shortness of breathe (SOB), stroke, seizures, diabetic reactions, shock, head injuries, etc.
- Using stabilizing techniques. Restrict movement of a vehicle in a collision by using stabilizing techniques.
- Re-set fire alarms and understand enunciator panels.
- Maneuver fire hose lines, handle nozzles and related hose line adapters and various types of portable extinguishers.
- Hook hoses to hydrants to create a water source for fire suppression activities.
- Understand sprinkler and standpipe systems, connections and other special fire protection systems.
- Maintain and operate illumination equipment, portable generators, gas and electric powered fans.
- Administer advanced first aid and perform artificial respiration and cardio-pulmonary resuscitation.
- Operate mechanical resuscitators and defibrillation machines.
- Enter burning structures to rescue and preserve life of occupants inside.
- Operate extinguishers, axes, pike poles, hydraulic tools, pneumatic tools, cutting torches and other equipment.
- Use various types of protective breathing apparatus when working in hazardous atmospheres.
- Work off ladders from various heights using an assortment of tools such as axes, pike poles, etc.
- Evaluate placement of ladders for safe and effective usage.
- Maintain and repair firefighting equipment.

- Understand and operate various types of detection instruments employed by the fire service to detect hazardous situations.
- Become thoroughly familiar with city streets, fire hydrants, intersection building numbers, building contents and occupancy.
- Maneuver charged hose lines into burning buildings for life rescue or fire extinguishment.
- Operate and drive fire apparatus transporting firefighters safely to and from emergency scenes
- Operate aerial ladders, integrated or portable pumps to provide fire streams and elevated work, and rescue platforms.
- Understand fire behavior such as back drafts, rollovers and flashovers.
- Continue learning in order to pass oral, practical or written examination mandated by department policy.
- Remain current in fire safety practices and fire safety awareness with students and faculties in the public and local schools.
- Respond to emergency call-backs while off duty.
- Respond immediately and safely to all emergency calls and requests for assistance.
- Maintain physical and mental fitness necessary to carry out all duties of a firefighter.
- Establish and maintain the confidence of the public.

CHAPTER TWO

HOW SOON SHOULD I START PLANNING?

High school is the ideal place to start planning and becoming enthusiastic about your future. It's where you form your learning skills towards an interesting lifetime career. It is also where you start to develop your life skills. Fire departments are interested in your entire background. They try to get a feel for the type of person you have become. If you have played many types of sports, you will more than likely be seen as a team player who is able to take orders and understand the importance of working with others properly. Are you someone who wants to be a firefighter down the road after you have achieved a university degree or a skilled trade?

How good does it look to a fire department when you were in high school that you were a volunteer in your community? A great deal! It shows the department that you have the caring attitude they're seeking.

It's never too soon to start planning for your future. Most courses offered in high school will benefit you in any career. However, fire-fighting is known to be a highly diversified profession and you never know what skills, old or new, you're going to rely on, and when.

Becoming a firefighter will not happen overnight. This career takes a great deal of patience and commitment. Sometimes it seems like you don't acquire a lot in return for courses you take or things you

You should start planning your future as soon as possible.

do along the way, but when you are hired on a fire department and receive the badge, everything comes into perspective. Every second spent to get there becomes important.

You must plan and become determined to accumulate all the prerequisites required in order for you to be eligible for most fire department recruitments over the period of a year. It may take another year if you decide to obtain your firefighter certification and take other fire-related courses. You may complete sooner if you schedule everything right. Be patient, it may take a little longer if the courses you need are offered on a limited basis. That's why the sooner you start the better will be your future opportunities.

It's critical to acquire the minimum education level. There is no point in thinking about getting hired on a fire department if you don't have the credentials. It is also good to have something to carry you, like a trade or a degree in case you are not immediately hired by a fire department. Be wise. Think ahead and have another career maintain a full life.

In order to discover new lands,
one must be willing to lose sight of the shore for a very long time.
— Anonymous

CHAPTER THREE

IS AGE A FACTOR?

Is age stopping you from pursuing a career in firefighting? Do you think you're too old or too young to get hired? These questions run through every potential firefighter's mind at some point. It's difficult to provide a straight answer to these questions due to the fact that there are many variables that affect the outcome. Fire departments don't have a specific age category, they prefer to hire from within. The main reason is simply because some of the best candidates recruited fall under the so called "too old or too young" category. Older or younger candidates have many qualities to offer fire departments, some are the same and some different. The disadvantage of hiring a younger firefighter recruit is his or her lack of life experience, but on the positive side; strength, endurance and impressionability is gained. If a mature firefighter recruit is hired, fire departments gain life experience, loyalty and work appreciation, but lose years of available service. It's stereotyping to assume every recruit falls under one of these categories. There are exceptions, but it's difficult for fire departments to determine what they are when departments are bombarded with over 2500 résumés.

If you're a young firefighter recruit and you're worried that you might not be hired because of your age, here is a tip for you: Fire departments assume because of your age you are lacking life experience as well as maturity and in the majority of cases a young fire recruit is only interested in the thrills of fighting fires and saving lives. Can you imagine how impressed an interview panel would be if you already understood the importance of fire prevention in the fire service? This would reflect the full potential you have to understand the roles and responsibilities of a firefighter as well as demonstrating your maturity level.

How do you act on this? The easiest way to get credit for having an interest in fire prevention is to have it on your résumé. Every fire department or the majority of fire departments have a fire prevention officer or fire prevention duties. If you show an interest that you would like to join the fire inspector to help out and gain some experience, they may allow you to ride along. This experience is added to your résumé under the heading 'Volunteer Experience'. Another positive view is the fire department will be able to put a face to your name, so the next time when they're recruiting firefighters it will be your name they recognize! It's really a win-win situation for you.

Just remember that fire departments are unpredictable and sometimes it's hard to figure out what their reasoning could be for not hiring you. That really depends on the structure of the fire department and their long-term goals. If you are still concerned or aren't sure if that fire department hires older or younger firefighters, you can research the ages of the firefighters that have been hired over the past five years and see their average age. You may find the answer surprising. Don't listen to others. I know for a fact where out of four firefighters recruited, one was eighteen years of age and another was forty-four. So, you see, age may work to your advantage and it's hard to say if age was even a factor. However, I would refrain from disclosing your age at any point during the hiring process, unless you are asked directly in an interview or on your application form. It's illegal to discriminate against someone because of age so most fire departments don't want to know how old you are until they decide to hire you. This way it doesn't affect their overall decision. The bottom line is to be someone who, through knowledge and interest, is valuable to the department.

Start by doing what's necessary, then what's possible,
and suddenly you are doing the impossible.
— Francis of Assisi

CHAPTER FOUR

VISITING YOUR PHYSICIAN

Before you spend all your time and money on courses it's important to first visit your physician. Make arrangements to have your doctor conduct a thorough physical examination to confirm you're in good health. You must meet the minimum physical health requirements requested by fire departments. You must have a minimum of 20/30 vision in both eyes without the use of corrective lenses. Laser surgery, if appropriate, can sometimes correct visual deficiencies. If laser surgery is not an option for you, it's best to find out prior to spending a significant amount of money on a firefighting school and fire related courses.

Fire departments expect you to be in excellent physical and mental condition and they don't accept anything less.

You must specifically inform your physician that you are thinking about a career in firefighting. This will ensure a thorough physical examination that is tailored to give you an accurate assessment of your physical condition. This way you eliminate any surprises later on in your recruitment process when you're undergoing a medical examination for the department that intends to hire you. Be sure to discuss with your physician about any medical concerns you may have; don't wait for a medical problem to be found especially allergies you could correct to increase your hiring chances. For instance, I played in a band for eight years and hearing loss was a concern of mine. So before I did anything I had an audiologist assess my hearing. Firefighting is an excellent reason to kick poor lifestyle habits such as smoking. You'll not only feel better, but others will take you more seriously when you say that you gave up smoking to become a firefighter.

Due to all of the physical testing as well as the physical strain endured on the job it's always wise to consult your doctor to make sure that your body and heart can handle the stresses of working out and getting into shape. Your doctor will be able to help you find a safe target heart rate that you can use as a guideline while working out. If you have not trained in a while or you have never done so before, be sure to take it easy for the first couple of weeks when you are at the gym. Your body will need time to adjust to the new level of strains and energy required. However, your body should adapt quickly. If for any reason you feel light headed or dizzy during or after a workout, be sure to consult your physician.

If you have any previous sports injuries or strains take the time to review it with your physician or fitness instructor to develop exercises you can perform to prevent the strain or injury from reoccurring.

Throughout your recruitment process and your firefighting career you are going to be demanding a great deal from your body. Be sure to take care of it.

"An ounce of prevention is worth a pound of cure"

CHAPTER FIVE

THE RIGHT ATTITUDE

Do you have what it takes to be a firefighter? No one knows the answer as well as you do. You're the one to decide your future. If you want to be a firefighter the only one who can make that happen is you.

Attitude plays a major role throughout your recruitment process. I can guarantee that you're in for a lot of ups and downs. The ups are really up and unfortunately the downs are really down. There is no middle ground. It's important that you enjoy and hold on to the positive things when they occur to get you through the negative times. Becoming a firefighter is not an easy task. The profession is highly sought after by individuals for a number of different reasons. Don't let anyone discourage you from pursuing your dream; be true to yourself but don't think it's going to be a walk in the park. Once you obtain all your prerequisites, including that positive attitude, fire departments are going to be knocking down your door offering you employment. It's important that you understand and develop the commitment level you need and the determination required to have a successful recruitment.

It's difficult to maintain the same level of intensity throughout your recruitment process. Through interaction with other candidates pursuing their dreams of being a firefighter, you can help each other stay focused and informed. By gaining insights about other candidates you may find invaluable information regarding excellent courses to take—possibly one you would have never thought of on your own. Perhaps you'll find it best not to be taking a particular course because it would have been a complete waste of time. These are things you need to know and evaluate because you can't afford to waste time on your journey to the career of a lifetime.

This job demands a great deal from you and that takes a lot of time, concentration, and energy. You may have to choose to miss important events in your life that conflict with you pursuing a firefighting career. But that's the type of dedication this job demands—and deserves; each sacrifice large or small is worth it. Have you wondered what the views were of other candidates regarding how much they wanted to be a firefighter? For instance, if you asked fifty candidates if there was anything they wouldn't do for a firefighting job, I guarantee you all fifty candidates would say "*No!*" So you have to ask yourself, 'Would I?' If your answer is "*Yes*," then you should rethink your career choice because you'll probably be discouraged by those who are very clear about how much they want to be a firefighter. It's important that you realize what you're up against so be prepared to be completely committed to your goal. If you believe in yourself and never stop believing, your dream will come true. Be positive in everything you do.

Remember to be nice to everyone—firefighters deal with every type of personality including city hall workers, city maintenance workers, hospitals, EMS, police, charitable organizations, schools, home and business owners during home inspections and the pre-planning of commercial establishments. It's important that you have excellent communication and interpersonal skills as well as a thorough understanding of how to treat people openly. Throughout your recruitment you're going to come into contact with many different individuals. With the proper attitude and the right approach it's possible to gain real supportive friends in high places. Being recruited onto a fire department is a little easier if you have a lot of reliable friends helping in your corner.

Courage is resistance to fear;
master of tear—not absence of fear.
— Mark Twain

So spend your time taking courses, making the right kind of new friends, and gathering information. Knowledge is the key, so never stop educating yourself.

❗ KEY POINTS TO REMEMBER

- 80% of emergency calls are medical related.
- There are benefits to being a young recruit as well as an older recruit.
- Have a doctor confirm your health.
- High school is a great place to start preparing yourself to be a firefighter.
- Never stop educating yourself.

NOTES

NOTES

Part II

Getting Ahead
of Your Competition

CHAPTER SIX

MINIMUM REQUIREMENTS FOR BECOMING A FIREFIGHTER

Every fire department you approach will have a list of minimum requirements that you must achieve prior to applying for a firefighting position. These minimum requirements will vary from department to department. Fire departments request that you present different requirements based on the types of calls they respond to. A fire department located by water may require you to have a certain water rescue course whereas a department located on an escarpment may require you to have your high angle rescue technician level. Be thorough when researching the minimum requirements for the fire department that interests you most. Many are similar so it means you will have to be qualified or eligible to work at a number of different departments. You should be able to determine the specific requirements by contacting the city's human resources department. You can also obtain the information from each fire department's website. Where you'll find a detailed outline of what is expected of firefighter candidates.

Be prepared; if you want to be a firefighter in the town where you grew up it's too late to find out their minimum requirements if you wait until they post a recruitment. Make sure the information you receive is accurate. You don't wait for the fire department to post a recruitment to find that you're not eligible to become a recruit.

EXAMPLE OF MINIMUM REQUIREMENTS:

- You must be legally entitled to work where you are living as a citizen, or landed immigrant.
- Proof of grade 12 education or equivalent.
- Meet the prescribed visual requirements of 20/30 in each eye without the use of corrective lenses and complete a colour vision assessment.
- Be able to communicate quickly in English.
- Copy of valid CPR and Standard First Aid Certificate Level-C.
- Copy of valid unrestricted Class 3 driver's licence with air brake endorsements.
- Copy of valid certification of successful completion of the fire fighting physical fitness test.
- Successfully complete an aptitude test and associated interview.

TIP!

If by chance you have a juvenile offence that you can't have removed from your record make sure you are upfront about it in an interview or if ever asked about it. The bottom line is if a fire department is interested in you they will find out about the offence, so you might as well be frank. This will also demonstrate that you accept your mistakes and that you are a trustworthy person.

Although the minimum qualifications are fairly easy to achieve, the fact is that career candidates pursuing their dream of a firefighting career raise the bar each time they apply.

These candidates have taken courses and achieved skills to improve their value on their résumé. In order to stay competitive you have to achieve the equivalent or better in order to remain a contender. If you have the right attitude this comment should not discourage you; just accept it as giving you an advantage by realizing that you can't waste time. Get going on your plan to take the courses that you know will best fill the gaps you found in your résumé or add strength to it. It is important to sit back to do an honest assessment of your résumé. What sense is there in taking another course that is similar to a course you already have passed if you haven't obtained all of your prerequisites? Your main priority is to be eligible to be hired by a fire department. Plan carefully, and then write it down to keep you on the right career path.

Fire departments will run a background test on you. There are ways to improve your background check and increase your chances of getting hired.

Take care of things that might draw attention to the background investigator:

CPR Level C, covers all aspects of CPR skills and the theory for adults, children and infants.

- Pay your old traffic tickets.
- Remove juvenile offenses from your record— the ones that your lawyer said wouldn't count.
- Have proof of full recovery after a serious illness.
- Drug free status since high school.

COMPARING MINIMUM REQUIREMENTS

ONTARIO Minimum Requirements	EASTERN & WESTERN CANADA Minimum Requirements	UNITED STATES Minimum Requirements
• High school diploma or equivalent • DZ driver's licence • First aid and CPR—level C • Vision 20/30—uncorrected • Valid physical fitness assessment • Successfully completed aptitude test and associated interview • Legally entitled to work (i.e., citizenship, landed immigrant or have a work permit) • Communicate in English **Recommended Qualification** • Post-Secondary fire program (Pre-service Firefighting Certificate, NFPA 1001) • EMR—Emergency Medical Responder	• NFPA 1001 Firefighter I & II • Medical and fitness assessment • Driver's abstract • Class 3 licence with valid air brake endorsement • First responder or industrial first aid or occupational first aid • Must be legally entitled to work (i.e., citizenship, a landed immigrant or have a work permit) • Understand and be able to communicate in English **NOTE:** Some fire departments require applicants to have both paramedic and NFPA—Firefighter Level I & II **Recommended Qualification** • Paramedic • EMT	• NFPA 1001 Firefighter I & II • Medical and fitness assessment • Driver's abstract • Class 3 licence with valid air brake endorsement • First responder or industrial first aid or occupational first aid • Must be legally entitled to work (i.e., citizenship, a landed immigrant or have a work permit) • Understand and be able to communicate in English **NOTE:** Some fire departments require applicants to have both paramedic and NFPA—Firefighter Level I & II **Recommended Qualification** • Paramedic • EMT

ACHIEVING MINIMUM REQUIREMENTS

Achieve the minimum requirements for your first choice of a fire department before achieving them for your second choice or third, especially if you know they're going to be recruiting in the near future. But you must also use common sense, obviously if your third choice is having a recruitment in the near future and your first and second choice have no intentions of having a recruitment then obtain all of the third choice minimum requirements first.

Even though you are focusing on three departments you still have to be willing to get hired anywhere.

What are the three fire departments where you prefer to be hired?

First Choice _____

Second Choice _____

Third Choice _____

There are a number of fire departments in your region, so it's impractical to have the minimum requirement for every fire department. However, you may find the minimum requirements for one fire department similar to others. Keep a good record in writing of all their needs.

What are the minimum requirements for each of the fire departments listed on page 26?

First Choice	Second Choice	Third Choice

It is also important to start learning about the fire departments you are pursuing: Try to find out what charities they support or are involved with. This will help you decide where to start volunteering and connecting. If one fire department raises money for children with learning disabilities wouldn't it make sense for you to volunteer with kids that have learning disabilities?

BRANDON MEETS HIS HERO AT STATION 245

When 6-year-old Brandon Quesnelle pulled a pot of boiling oil on top of himself in a cooking accident almost two months ago, it was Toronto firefighters who tended to his injuries and started him on the long road to recovery.

Yesterday the youngster re-established the bond that was forged in that moment of intense pain and fear, visiting the firefighters at Scarborough Station 245 at Birchmount Rd. and Ellesmere Rd.

The blue-eyed boy in a blue polo shirt stepped out of his mother's van at the station.

He was greeted by firefighter Brian Fogarty, who treated him in the ambulance after the scalding oil ran down his shoulder and back, causing severe burns. Brandon spent the next five weeks in hospital and was only able to return to school on March 27.

"Do you remember the ambulance ride?" Fogarty said.

"No," said Brandon, who then sat on the ground and crossed his legs. "Just my leg hurts when I do this."

Then the little boy wanted to see the fire truck and slide down the pole and try on their boots.

His eyes lit up when he saw the presents from his heroes sitting on the bumper of the truck. The package include a Tonka fire engine, a hockey stick and a Scottish national soccer team jersey.

He also received a package from Camp BUCKO—Burn Camp for Kids in Ontario—sponsored by the Toronto Firefighters Association. Brandon can go for one week each summer for 10 years.

"We don't really get many good moments," Fogarty said after the meeting. "He looks a lot better than I had expected."

— By: Himani Ediriweera. *Toronto Sun.* April 3, 2006

CHAPTER SEVEN

CHOOSING THE RIGHT COURSES

Through diversified training and continued education firefighters have equipped themselves to handle any type of situation. No matter the call of duty—a motor vehicle collision, medical emergency, hazardous materials incident or a fire; firefighters are ready and able to handle it. Firefighters also need general knowledge in maintaining and using stand pipe and sprinkler systems as well as possess an in-depth knowledge of building construction. In order for you to be well qualified it will take many hours and substantial expenses. If you're just starting to pursue a career in firefighting you must face competition from other candidates who have been pursuing their dreams of becoming a firefighter, too.

So how are you supposed to measure up to become a contender with these candidates? It's not easy but it's definitely possible as candidates like you do it all the time. Through careful planning and strategic course selection you can find yourself being a serious contender within a reasonable amount of time. It really depends on how motivated you are as well as the amount of time you can dedicate to this process. By choosing what I call "cookie cutter" courses you're obtaining courses that most everyone has acquired. This is exactly what you don't want. You need to be different—achieve courses that will set you apart from everyone else. You want whoever is reviewing

Offer the fire departments your full potential. Help them create depth in their fire personnel.

your achievements to be impressed with how hard you've worked and studied to achieve the courses you have on your résumé. You must make it easy to be pictured as someone who should be working in their fire department. If you have what the fire department is searching for, then chances are you're going to get that first interview.

Too many candidates take the attitude; if I'm lucky I'll get that first interview. Luck has nothing to do with it.

Whenever I asked myself, 'Would I hire me?' I didn't stop completing courses and improving myself until I could honestly answer 'Yes'.

To help you decide what courses to complete I have compiled the following list. These courses may not guarantee you an interview, but they will keep you on the right track and miles ahead of those with résumés that look like everyone else's.

TIP!

An easy course to obtain in a weekend that is often overlooked by candidates is the certification to install car seats. Many fire departments are involved in car seat installation clinics that help the public install their car seats properly. These courses are usually available through city's health units, if not contact your local fire department and see if they can help you locate a course provider.

Courses you should consider:

Fire Training
- EST (Emergency Medical Technician) Program
- Pre-service firefighter certificate
- SCUBA licence
- Pump "B" operator
- Silo fire awareness
- Fire prevention/investigation division
- Fire arson investigation
- Scott air pack field-level maintenance, technician
- Marine firefighting for land-based firefighting
- SCBA repair
- NFPA Courses:
NFPA 1000
  Standard for Fire Service Professional Qualifications Accreditation and Certification Systems
NFPA 1001
  Standard for Fire Fighter Professional Qualifications

NFPA 1002
  Standard for Fire Apparatus Driver/Operator Professional Qualifications
NFPA 1003
  Standard for Airport Fire Fighter Professional Qualifications
NFPA 1005
Standard on Professional Qualifications for Marine Fire Fighting for Land-Based
  Fire Fighters
NFPA 1006
  Standard for Rescue Technician Professional Qualifications
NFPA 1021
  Standard for Fire Officer Professional Qualifications
NFPA 1031
  Standard for Professional Qualifications for Fire Inspector and Plan Examiner
NFPA 1033
  Standard for Professional Qualifications for Fire Investigator
NFPA 1035
  Standard for Professional Qualifications for Public Fire and Life Safety Educator
NFPA 1037
  Standard for Professional Qualifications for Fire Marshals
NFPA 1041
  Standard for Fire Service Instructor Professional Qualifications
NFPA 1051
  Standard for Wildland Fire Fighter Professional Qualifications
NFPA 1061
  Standard for Professional Qualifications for Public Safety Telecommunicator
NFPA 1071
  Standard for Emergency Vehicle Technician Professional Qualifications
NFPA 1081
  Standard for Industrial Fire Brigade Member Professional Qualifications

Rescue
- Bronze cross
- Bronze medallion
- Confined space rescue
- Farm accident rescue

- Hazardous materials-awareness, operations or technician level
- Ice rescue
- National life guard qualification
- Rope rescue
- River rescue
- Swift water rescue

Medical
- AED (Automated External Defibrillator)
- Air ambulance landing zone preparation
- BTLS (Basic Trauma Life Support)
- EMR (Emergency Medical Responder)
- First aid instructor
- Intermediate cardiac life support
- Paramedic
- Pediatric Pre-hospital Care

Operators
- Aerial work platforms
- Class 1 or class "A" licence
- Class 2 or class "B" licence
- Class 3 or class "D" licence
- Class "C" licence
- Defensive driving
- Industrial truck operator
- International hand signs (Rigging)
- Pleasure craft operator licence

Communication
- Sign language
- Emergency services communicator certificate
- Public speaking experience
- Crisis intervention
- Emergency medical dispatcher
- Restricted radio operations licence

General

- Administration courses
- Any trade: electrician, plumber, refrigeration
- Blueprint reading
- Building construction
- Carbon Monoxide
- Car seat Installer (baby seats)
- Critical incident stress management
- Energy control & power lockout
- Fall arrest certificate
- Fire alarms
- Gas technician I, II, or III
- Health & safety committee
- Outdoor recreational leadership diploma
- Service & repair small engines
- Sprinkler system
- Transportation of dangerous goods

The best way to obtain the courses you need is quite simply to approach the situation the same way I talked about approaching your minimum requirements. Focus on three departments and make sure you take the courses that will benefit the fire department where you first wish to be hired. Then take other courses that would be good to have (see other department's requirements). For example, with the three fire departments that you choose, find out specifically the limitations of each department regarding emergency calls. If one of the three require high angle rescue emergencies, but all three do water rescue emergencies, it would only make sense to get your water rescue certification before the high angle rescue. To assure your time is well spent, you must prioritize every course until you have obtained everyone that is a prerequisite. Don't worry about taking specialty courses before you have all of your prerequisite courses. Your main concern is to be eligible when fire department recruitments occur. You may even be hired before you spend a lot of money on non prerequisite courses—but never lose your edge for learning.

Course Brainstorming

Throughout your recruitment process you will discover that you require more courses/certificates in order to remain a competitive contender. Perhaps a course is recommended to you. But by the time you get to the point where you can consider taking the course you have already forgotten what it was or where it was offered. This is very common because there are so many dates and times you must remember that it's impossible to keep track of everything.

Whenever you come across a course or certification you find valuable to your résumé write it down immediately as part of future plans. Why lose track of a new lead that will benefit your career?

The course brainstorm chart that follows will allow you to record courses and certifications you find and provide the opportunity to research them when you have spare time. Once you have them researched transfer them into the respective category "short term course" or "long term course".

Short term course — These are courses that you can obtain with limited time and money. Usually these are one day or weekend courses which are offered frequently.

Long term course — They are more involved and usually require a great deal of commitment. Planning is needed to make arrangements for you to attend.

Choose a course from your short term list when time and money are limited. If you anticipate an increase in time and in the near future you may consider achieving a course from your long term course list. By using this system you are always prepared and obtaining the qualifications that are going to put you in front of your competition.

People with goals succeed because they know where they're going.
— Earl Nightingale

Figure 2-1
Course brainstorm chart

Course Brainstorm

1.	15.
2.	16.
3.	17.
4.	18.
5.	19.
6.	20.
7.	21.
8.	22.
9.	23.
10.	24.
11.	25.
12.	26.
13.	27.
14.	28.

Short Term Courses | Long Term Courses

Short Term Courses	Long Term Courses
1.	1.
2.	2.
3.	3.
4.	4.
5.	5.
6.	6.
7.	7.
8.	8.
9.	9.
10.	10.
11.	11.
12.	12.
13.	13.
14.	14.

COURSE MANAGER

Take advantage of this organizing tool. The "Course Manager" chart allows you to keep a log of the past and upcoming courses taken or you want to take.

Course	Date	Length of Course	Location	Date Course Completed	Date Added To Resume

There are many different ways to get a job at a fire department and more than one path to take. I want to teach you how to decide what path is best for you personally. For example, a First Aid and CPR instructor's certificate would be an excellent qualification to have on any résumé. Personally, I didn't feel it was a route that best suited me. Even though being a First Aid and CPR instructor would be an impressive addition to my résumé, as well as a benefit for any fire department, I still felt it was a course I was going to postpone for the time being. Some of you may feel the opposite about this and that's fine; you have to take the courses that interest and stimulate you and are going to make you an asset to the fire department you may want to join.

Just remember to make every minute of every day count. Make an effort each day, whether it's visiting a fire hall or reading a newspaper article about a recent fire in your hometown; or even cut out articles and start a scrapbook. Many ways exist to keep your head in the game. Again, find out what works best for you. Having a one track mind can be a big benefit!

Goals are dreams with deadlines.
— Diana Scharf Hunt

Medical Experience Overview

If you want to be a firefighter you must obtain a minimum level of medical training. Depending on where you live will determine the level of medical training fire departments require that you have prior to being offered employment. Be sure to pay close attention to this because fire department protocols are changing constantly, so be sure to confirm to the prerequisite medical training periodically.

Times have changed. Firefighters don't respond to only fires. Most career fire departments respond to medical emergencies and continually trained to a first medical responder qualification level. Most don't realize that 80% of calls are medical-related emergencies. Most fire departments are on the scene within 4 to 5 minutes of receiving the alarm and often before an ambulance. In many cases, this is often the difference between life and death.

Each departmental region has different policies and guidelines regarding the minimum requirements for each level of emergency medical certification. Some don't recognize certain levels of emergency medical training. For example, Ontario is a province that doesn't recognize EMT-B (Emergency Medical Training-Basic) but for most provinces in Canada, EMT-B is a prerequisite for any candidate applying for a firefighting position.

Before you spend a lot of time, money, and energy in taking specific medical courses be sure to research what regions recognize that type of training. However, if you've already obtained a medical course certificate that isn't reconized in a specific province, chances are you will still get credit for the training even though you won't be able to use it. This means that you can't perform to the level in which you've been trained, so in this case the course may have got you the job even though you can't completely use your skills.

If you're applying to a fire department that only requires you to possess a minimum of basic first aid and CPR-level "C" then any medical training you receive beyond that will be a bonus to you and your résumé. Unfortunately for those fire departments that require higher levels of medical training such as EMT-B or Paramedic as a prerequisite, you may have to find other areas on your résumé to pick up bonus points. No matter what level of medical training required, the common denominator is EXPERIENCE! You can have all kinds of medical certificates on your résumé but the candidates with actual experience will get the job every time.

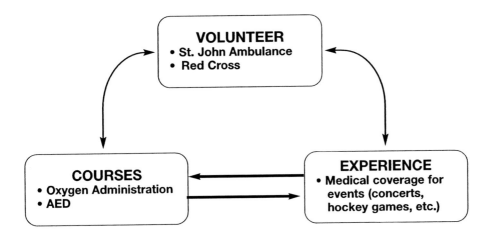

My advice for you is to get involved volunteering with a medical response or medical related non-profit organization. You will accomplish two objectives simultaneously: You will build your volunteer experience for your résumé; and you will gain confidence as well as experience handling medical non-emergency and emergency situations. Having this experience will greatly improve the depth of your résumé as well as assist you in your interview when asked questions, "Name a time when you had to respond quickly and efficiently to handle an emergency situation?" You should be able to reflect back to a date, time and involvement in the actual events that had happened to you while you were volunteering. Being able to give a detailed answer to this question will greatly improve the outcome of your interview.

During your recruitment process you will find that everything comes back to selecting the right courses that give you the edge over competition. There are several levels of medical training you can achieve as well as a number of pre-hospital emergency courses. Be sure to be creative and selective when choosing medical courses. Remember to have diversity in your training and try to accumulate as much actual medical experience as you can.

FIREFIGHTER GOES BEYOND THE CALL OF DUTY

Saving lives is in a firefighter's job description. Seth Wells figured he would just go about it in a different way.

The 27-year-old civilian firefighter at Vandenberg Air Force Base returned home from Boston one weekend after donating his kidney to a fellow firefighter he had briefly met. Wells gave a kidney to Walter Stecchi, 45, who worked at Vandenberg in the 1980s and is currently an assistant chief at Otis Air Force Base, Mass.

Stecchi's kidneys had been going downhill for nearly 20 years and his body had already rejected a kidney from his father. Stecchi was on dialysis when his firefighters union sent a letter to other bases requesting donors. Wells immediately stepped forward, something Stecchi wasn't expecting.

"I was happy, surprised. I thought it was pretty amazing someone would step up like that. I didn't have much hope in the beginning," Stecchi said. Wells first met Stecchi when he flew to Boston in September to undergo testing. He met him a second time last week, before surgeons performed the surgery.

The transplant was an immediate success and the effect on Stecchi was like night and day, according to both men. After leaving the hospital, Wells and his family spent the week at Stecchi's home before returning to the Central Coast Friday night. Wells' colleagues from Vandenberg Fire Department gathered at the airport in Santa Barbara to greet him, his wife and child upon their return.

Wells is eligible for 30 days off and says Vandenberg officials have told him to take as much time as he needs. Stecchi's insurance paid the hospital bills and Well's firefighter union paid for his travel expenses.

"A lot of guys think I'm nuts. It's not about what if I need the kidney. It's what if I don't need it?" Wells said from his Grover Beach home where he is recuperating. "We're in this career field where we're trying to help people so I try to do that to the fullest extent."

Stecchi says he considers Wells a brother and both plan to stay in touch. "It's nice to have another set of family," Wells said. "We're going to meet every year, just hang out. It'll be a bond for many years to come."

— By Mark Baylis, www.firerescue1.com

CHAPTER EIGHT

CHOOSING A FIREFIGHTING PROGRAM

You want to attend fire school but you're not sure what course to take? Or what school to attend?
Finding those answers can be difficult. Ask ten different people and you get ten different answers and everyone would be right one way or another. Its one thing to find a good fire school, but it's another to find the right fire school. Choosing the perfect fire school program will depend on such criteria as where you live geographically, because different regions endorse different fire programs or schools. For example, if you would prefer to be hired by a fire department on the west coast then it's recommended you attend a fire school in that area.

You also have to consider your level of commitment and your family and financial situation. If you're someone who has obligations to a young family as well as paying down a mortgage then realistically it may not be practical to enroll in a three year, full time post secondary fire program. You may have to consider taking a condensed program, correspondence program, or complete your course over an extended period of time. On the other hand, if you're someone fresh out of high school, then taking a longer post secondary fire program will allow time for you to mature and gain valuable life experience. Both of these credentials will benefit you down the road in your pursuit to being a firefighter. You'll have to make sacrifices; some big and some small. It's having the attitude to make these sacrifices that will help you widen the gap between you and your competition.

When choosing a fire school or program questions must be asked: Is the program recognized by the area you want to work in? Does the program meet the proper standards? The last thing you want is to spend your time and money on a fire program that isn't recognized or accredited. Governments and organizations regulate the standards of fire training and accredit entities in which to teach the standard. These regulations are in place to ensure that you receive a level of training that gives you a strong foundation in which to start your firefighting career.

There are three abbreviated organizations you'll see throughout your recruitment process and fire career. It's important that you understand the relationship between them and their significance when choosing a fire program:

NFPA   (National Fire Protection Association) is an international nonprofit membership organization founded in 1896. NFPA serves as the world's leading advocate of fire prevention and is an authoritative source on public safety. NFPA's 300 codes and standards influence every building, process, service, design, and installation, as well as many of those used in numerous countries. NFPA is driven by more than 6000 volunteers from diverse professional backgrounds who serve on 230 technical code and standard-development committees.

Motivation is what gets you started.
Habit is what keeps you going.
— Jim Ryun

IFSAC (International Fire Service Accreditation Congress) is a peer driven, self governing system that accredits both fire service certification programs and higher education fire-related degree programs. IFSAC is a nonprofit organization authorized by the Board of Regents of Oklahoma State University as a part of the College of Engineering, Architecture and Technology. The IFSAC Administrative Offices are located on the Oklahoma State University campus in Stillwater, Oklahoma.

IFSTA (International Fire Service Training Association) was established as a nonprofit educational association of fire fighting techniques and safety through training. This training association began in 1934. Fire Protection Publications' primary function was to publish and disseminate training texts as proposed and validated by IFSTA. As a secondary function, Fire Protection Publications researches, acquires, produces, and markets high-quality learning and teaching aids as consistent with IFSTA's mission. IFSTA's purpose is to validate training materials for publication, develop training materials for publication, edit proposed rough drafts, add new techniques and developments, and delete obsolete and outmoded methods.

Figure 2-3

In order to fully understand the relationship between NFPA, IFSAC and IFSTA use this flowchart.

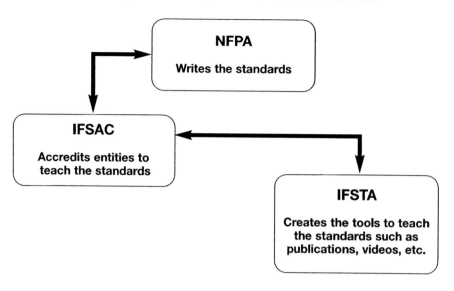

The best way to look at a comparison of the three organizations is that NFPA develops the standards, IFSTA develops training materials based on the NFPA standards, and IFSAC accredits those organizations that certify firefighters based on NFPA standards as well as those institutions that offer fire-related degree programs. Ontario however has its own standard of firefighting training. The OFM (Ontario Fire Marshal) has created its own curriculum that is recognized by some fire departments in Ontario but not by fire departments outside of Ontario. So keep in mind, if you enroll in a course out of your home province be sure that it will be recognized by the fire departments where you want to be hired.

Now that you understand the standards and the powers that govern the pre-fire service training you have to decide what school you're going to attend if you're eligible. There are a number of schools that offer firefighting programs. First, you must decide the type of firefighter you want to be: structural, wildland, industrial, etc. Once you have decided that you need to research the schools that offer programs in that specific field. Firefighting schools have come a long way in the last few years and the quality of training is standardized. As long as the school is recognized by your government or accredited by NFPA you're on the right track. Talk to graduates of the fire school where planning to enroll to and ask what they thought of their experiences and the expertise of the school's instructors and curriculum.

SOME QUESTIONS TO ASK:

What criteria must be obtained prior to attending?
- Physical fitness test—which one?
- Level of education—chemistry, biology, grade 12 diploma?
- Language
- High school transcripts

General questions to ask about the school.
- Is the school accredited or recognized?
- Does the province you want to work in recognize the school's fire programs?
- What qualifications will you receive upon completion of the fire programs?
- How many students were in each class?
- What is the ratio between students and instructors?
- How much practical training given?
- How much did they prepare you for the real world? (Get details of what they did)
- Is the school setting relaxed or very structured like the military?
- Is extra help available (like mentoring) if you need it?
- Is there a recreational facility available?
- Are courses part time or through correspondence?
- Does the school meet today's standards?
- Is the equipment in good condition and relevant regarding training props, apparatus, training techniques?

Costs to consider:
- Tuition
- Books and supplies
- Admission fees
- Residence
- Parking

- Meals—cafeteria, restaurants, personal fridge space etc.
- Room and board (if available)
- Transportation—taxi, bus, etc.
- How much did you spend in total while attending school?

Obtaining and recording answers to these questions will make your decision easier. Use following school selection guide (on page 47) to help you decide the best course for you.

SCHOOL SELECTION GUIDE

Choose Three Schools You Would Prefer to Attend:

FIRST CHOICE	**SECOND CHOICE**	**THIRD CHOICE**
_____	_____	_____
Name of School:	Name of School:	Name of School:
_____	_____	_____
Cost of Tuition:	Cost of Tuition:	Cost of Tuition:

Location:

FIRST CHOICE	**SECOND CHOICE**	**THIRD CHOICE**
Address:	Address:	Address:
_____	_____	_____
_____	_____	_____
_____	_____	_____
Phone:	Phone:	Phone:
_____	_____	_____
Fax:	Fax:	Fax:
_____	_____	_____
Email:	Email:	Email:
_____	_____	_____
Website:	Website:	Website

COURSE PROFILE:

FIRST CHOICE	SECOND CHOICE	THIRD CHOICE
_____	_____	_____
Program Name:	Program Name:	Program Name:
_____	_____	_____
Length of Program	Length of Program	Length of Program

How many times a year is the course offered?

First Choice:
- ☐ Once
- ☐ Twice
- ☐ Multiple

Second Choice
- ☐ Once
- ☐ Twice
- ☐ Multiple

Third Choice
- ☐ Once
- ☐ Twice
- ☐ Multiple

How long is the waiting list?

FIRST CHOICE	SECOND CHOICE	THIRD CHOICE
_____	_____	_____
_____	_____	_____

Is the course accredited?

	Yes	No
First Choice	☐	☐
Second Choice	☐	☐
Third Choice	☐	☐

Is the course recognized by the fire departments where you want to be hired?

	Yes	No
First Choice	☐	☐
Second Choice	☐	☐
Third Choice	☐	☐

What certificates do you receive on course completion?

FIRST CHOICE	SECOND CHOICE	THIRD CHOICE
_____	_____	_____
_____	_____	_____
_____	_____	_____

Prerequisites for the course:

FIRST CHOICE	SECOND CHOICE	THIRD CHOICE
_____	_____	_____
_____	_____	_____
_____	_____	_____

Prerequisites for the course (continued)

FIRST CHOICE	SECOND CHOICE	THIRD CHOICE

CHAPTER NINE

VOLUNTEERING

How much does volunteering matter on my résumé? How much volunteering is too much?

Two very good questions with no exact answer. What I can say is without any volunteer experience on your résumé you will be at a big disadvantage. Many fire departments look at volunteering as a symbol of how much you care about your community. Since fire departments are well known for their community involvement they like to see the same from their potential recruits. It's more rewarding to fire departments to recruit individuals who have selflessly donated their time to assist others in need or those who are less fortunate. Volunteering demonstrates great characteristics of a firefighter and I recommend you take it seriously. It's hard to say exactly how much of your time you need to dedicate to volunteering but make your community part of your life. There is no set number of hours one must donate. Some fire departments or cities don't really put a lot of emphasis on volunteering, while others factor it in when deciding who they're going to hire. Since you're on the outside looking in it's really difficult to determine the importance of volunteering to each individual fire department or city. What you want to do is cover the basics and make sure you can demonstrate that you have made a worthy effort and are concerned.

When volunteering, try to do "what you can, when you can."

Where you decide to volunteer is completely up to you. In fact, most cities don't have a preference as to where you should volunteer. What they do want to see is an ongoing commitment to your city or town. For example, volunteering for two hours at a food drive one day last year won't compare to someone who has volunteered two hours a month for the last year and a half at the local hospital. (Keep in mind most hospitals have volunteer opportunities available, although duties may vary from hospital to hospital you may have to investigate every hospital in your area and determine which one offers you the greatest amount of experience. If possible get involved in the emergency departments to receive invaluable hands-on training that could cross over to situations found in the emergency services.)

Volunteering has many benefits, one being the fact that it's a great opportunity to meet other individuals who may also be pursuing firefighting as a career. It's important you make friends with these individuals because believe it or not as a team player you can help each other. It's impossible to be aware of everything that's happening with recruitments and courses being offered. By making friends it greatly improves your knowledge about what's going on.

There may also be monetary advantages to volunteering that you may not have considered. If you choose your volunteering placements wisely you may find that some volunteer organizations require its volunteers to have specific training. Chances are the training will be supplied free of charge or at a reduced rate. Three things are happening here; you're gaining volunteer experience, you're obtaining courses at a reduced price and adding them to your résumé and you're helping your community.

Research the top three fire departments of choice, then get to know the programs or organizations in which they are involved. If the chief of one of your chosen fire departments is involved with children with autism, you can't go wrong to volunteer for any organizations that raise money for autism, preferably in the city where you are planning to be hired. Now you share a common interest with the chief on an issue he feels strongly about—a definite feather in your hat. You should be able to find this type of connection in most fire departments, but it's up to you to find out what it is.

Another way to gain invaluable volunteer experience is to assist local fire departments with their own fund raising events. It takes a number of volunteers to run and organize a successful event, so usually fire departments appreciate any extra help being offered. What a great way to get your foot in the door and put a face to your name.

High achievement always takes place in the framework of high expectation.
— Jack Kinder

❗ KEY POINTS TO REMEMBER

- Minimum requirements may vary from fire department to fire department.
- Choose courses that set you apart from your competition.
- When choosing a fire fighting program be sure to choose one that's recognized by the fire department where you wish to get hired.
- Volunteer experience is as equally important as courses on your résumé.

NOTES

_____

_____

_____

_____

_____

_____

_____

_____

_____

_____

_____

_____

_____

_____

_____

NOTES

NOTES

Part III

Recruitment Process
Strategies

CHAPTER TEN

MUNICIPAL FIREFIGHTER RECRUITMENT PROCESS

Fire department recruitment processes vary from department to department. Each fire department has the authority to establish what measures are taken to find the best applicants and what minimum qualifications and standards are required in order to be hired. Recruitment processes can be complicated and

involved or short and simple. The objective of a fire department recruitment process is to determine an applicant's aptitude towards the profession as a fire fighter. Fire departments use multiple evaluations to determine each applicant's ability to execute fire fighting related tasks, ability to function under physical exertion, determine the overall health and fitness level of each applicant as well as the ability to understand written and verbal information.

Fire departments are unpredictable. It is difficult to determine when exactly a fire department will announce recruitment. You will hear rumors that a fire department is hiring long before a recruitment is posted. To be a successful recruit you have to be prepared for anything. Sometimes multiple fire departments recruit at the same time and sometimes there are none. You must be patient and maintain a positive attitude while waiting for opportunities to arise.

In order to outline the different stages of a firefighter recruitment process, fire departments use recruitment programs. These recruitment programs help determine which candidates possess the skills, aptitude, attitude and abilities needed to be a successful firefighter. Potential firefighter candidates must be successful in each stage of the recruitment program in order to be considered for employment.

COMMON STAGES OF RECRUITMENT PROGRAMS

- Posting of recruitment
- Meet the application submission deadline
- Ensuring applicant meets minimum qualifications
- Written aptitude test
- Notification of test results
- Process of scoring applications
- Notifying applicants of first interview
- Notifying applicants of second interview
- Arranging physical ability test
- Medical examination
- Confirming applicant passes both tests physical ability and medical
- References contacted
- Conditional job offer made

If you have been informed that a fire department is recruiting by someone who is not the definitive source, be sure to locate the information yourself.

Example: Posting of Recruitment

PROBATIONARY FIREFIGHTER RECRUITMENT   July 18, 2006

The Winnipeg Fire Paramedic Services is conducting a recruitment process effective 5 November 2005, with a closing deadline of 4:30 p.m. on 19 December 2005. The PDF document entitled Consolidated Recruitment Documents contains all the printable forms and requirements that candidates are required to fill out or provide. Please print and complete them exactly as indicated. WFPS thanks all applicants, but only those whose applications were deemed suitable for progressing to the testing and interview stages will be contacted.

QUALIFICATIONS:

1. Grade 12, according to Canadian Provincial Standards, GED or equivalent.
2. CMA Accredited Paramedic Program (minimum PCP)
3. Possess a valid Provincial License or be able to acquire
4. IFSAC or Pro board Accredited NFPA 1001 Firefighter Level II program
5. Current CPR Certification
6. Clear Criminal Record check
7. Valid Class 4 Driver's License or equivalent with no more than 4 demerits and no alcohol related charges / convictions for the last 4 years and a Drivers Abstract
8. CPAT Certification (Candidate Physical Ability Test) or a certified job related physical fitness test acceptable to the Service completed within the past 12 months.
9. The ability to successfully complete the EMS and fire Rescue Aptitude assessments, EMS and fire scenarios and a driving evaluation.
10. Ability to undergo and pass a medical examination administered by

Submit to: Human Resources
Fire Paramedic Service
2nd Floor, 185 King Street
Winnipeg, Manitoba R3B 1J1
(204)986-6369 www.winnipeg.ca

Once the decision is made by the municipality and fire department to recruit additional firefighters the first step in this process is releasing a public announcement. This is accomplished by placing an advertisement, or sending a press release by various medias such as local newspapers and the internet. Candidates interested in applying for the fire department must pay close attention to details in any article regarding the recruitment. The article will have critical information such as where to obtain an application form, application deadline and where the application is to be submitted, etc.

Candidates must confirm any third party information pertaining to any fire department recruitment. It is in the best interest of the candidate to contact a definite source regarding the potential recruitment. Don't rely on what people say to be completely accurate, you may find out the hard way that their information was misleading.

Meet the application submission deadline

No exceptions are made for late application submissions or failure to pay the entrance fee. Every fire department discloses application submission deadlines; be sure you understand when it is. Do not submit your application at the last minute. Failure to submit on time convinces the fire department that you cannot follow instructions and you lack self discipline. Your application will not be accepted.

Note: Applicants should notify the Human Resources Department or the fire department in writing of any additional skills and education acquired since your application. Copies of certificates, diplomas etc. should accompany the notification. This new information will be added to your personal file with the fire department and will be taken into consideration upon résumé evaluation.

First say to yourself what you would be;
and then do what you have to do.
— Epictetus

Ensuring applicant meets minimum qualifications

Fire departments insist on confirming every qualification reported on your résumé. This assures the quality of training each applicant has received.

Written aptitude test

In order to advance in the recruitment process applicants must successfully pass the written aptitude test presented by the fire department. Tests rely heavily on written aptitude as a way to ensure the quality of the applicants.

Notification of test results

Applicants can expect to receive their test results within a relatively short time after completion. Results may take longer depending on the volume of applicants. Typically test results are mailed to applicants, notification of whether or not you advance in the recruitment process are disclosed in this letter.

Unsuccessful applicants will simply receive a letter from the fire department thanking the applicant for their interest in the fire department that advises that they will not be able to continue in the recruitment process.

Process of scoring applications

Successful résumés will be assessed at this time by the fire department to determine what applicants are worthy of receiving an interview.

Notifying applicants of first interview

Purpose of an interview is not to outline the applicant's technical knowledge, but to determine an applicant's employment experience, related skills, and volunteer experience as well as enable the board an opportunity to further assess the applicants' personality and interpersonal skills.

Failure to meet any one of the minimum requirements means your application will be moved to a "failed" file.

Other details especially why the applicant wants to be a firefighter and why the applicant feels they should be considered over other qualified applicants are considered as well.

Fire departments typically notify successful first round interview applicants of a second interview by telephone. If the offer is accepted by the applicant, the time, date as well as location of the interview will be provided at this time.

Notifying applicants of second interview

If your first interview was successful you may be asked to return for a second. Fire departments may use two or more interviews to filter out ideal candidates. This is more common when the number of applicants is high.

Arrangements made for physical ability test

Once the applicant has successfully passed the interview stage the next hurdle is the physical fitness test.

Medical examinations

It is mandatory to have a thorough medical examination completed prior to being offered employment by the fire department. Any health concerns or issues with applicants are revealed during this hurdle.

Confirming applicant passes both tests physical ability and medical

Fire departments must have written documentation of both physical and medical assessments to confirm you are fit for duty.

References contacted

A complete background search and reference check is completed by the recruiting fire department. It is in the best interest of the applicant to disclose any information that may concern the fire department; honesty is vital to your success.

**Applicants failing to meet
any of the following requirements may be
disqualified from the recruitment.**

- Intentionally supplying inaccurate or false information about yourself.

- Failure to submit a complete application form.

- Failure to submit documentation supporting your qualifications on your résumé.

- Failure to submit required documentation by the application deadline.

- Failure to submit recruitment entrance fee.

Conditional job offer made

Once all the above criteria are met the fire department should be convinced you are a worthy candidate. Usually job offers are conditional because unknown facts such as: does the applicant still desire to work for the fire department that is presenting the job offer. Sometimes candidates receive more than one job offer at the same time, so obviously the candidate has to decide where they want to work. Once the job offer is accepted the applicant will have a verbal acceptance agreement with the fire department. It will remain a verbal by both parties acceptance agreement until all necessary legal documents are signed.

Congratulations you are now officially a probationary firefighter!

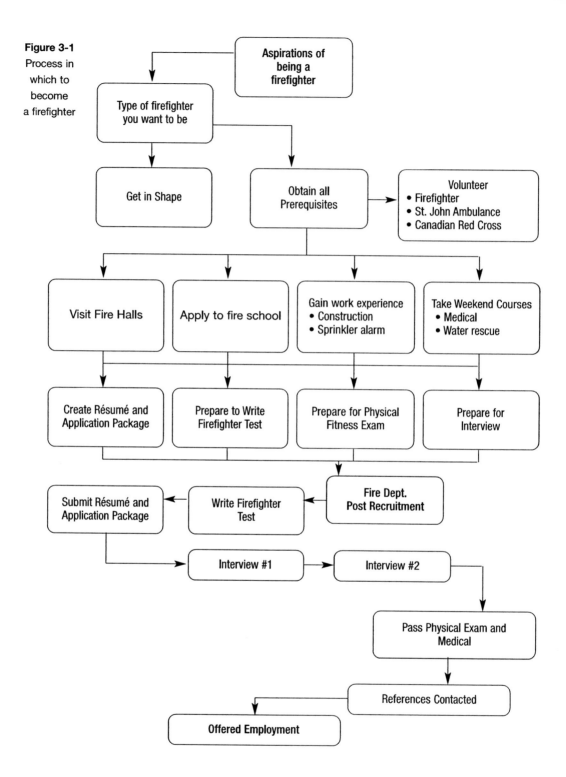

Figure 3-1
Process in which to become a firefighter

Aspirations of being a firefighter

Type of firefighter you want to be

Get in Shape

Obtain all Prerequisites

Volunteer
• Firefighter
• St. John Ambulance
• Canadian Red Cross

Visit Fire Halls

Apply to fire school

Gain work experience
• Construction
• Sprinkler alarm

Take Weekend Courses
• Medical
• Water rescue

Create Résumé and Application Package

Prepare to Write Firefighter Test

Prepare for Physical Fitness Exam

Prepare for Interview

Fire Dept. Post Recruitment

Submit Résumé and Application Package

Write Firefighter Test

Interview #1

Interview #2

Pass Physical Exam and Medical

References Contacted

Offered Employment

The process used to select applicants is a rigorous one. Don't be discouraged or surprised if you don't succeed the first time. It is natural to be disappointed in yourself or the process, but you can't let it discourage you. Anything worth doing doesn't come easy. There are a lot of steps to take in becoming a firefighter. Don't be surprised if you are unsuccessful in a stage or need some work in a specific area. Try to learn from your mistakes and be sure to build knowledge and confidence from them. Here are some helpful tips to help confront this misfortune.

IF YOU ARE NOT SUCCESSFUL THE FIRST TIME

Written Aptitude Test: Applicants can usually re-test after a waiting period. During the waiting period take the time to determine the areas of difficulty and prepare yourself to do better on the next test. Some fire departments use subcategories throughout the written aptitude tests. This will help you identify the areas where you require assistance to get a higher mark.

Physical Fitness Test: If you were unable to complete a component of the fitness test be sure to incorporate exercises in your workout that will help you successfully pass that component the next time.

Interview: Find out where you went wrong. If nerves were a problem, and you feel you didn't answer the questions as well as you could, use mock-interviews with friends or family, or a friendly firefighter to help you become accustomed to interview settings. Perhaps you were unprepared and thought you would just 'wing it', instead of reviewing your answers.

Medical Exam: Be sure to find out why you were unsuccessful and investigate what your options are. If necessary consult your physician (and a trainer) for advice.

Discrimination: If at any time you felt you were treated unfairly, for any reason such as racial discrimination, wrong attitude taken by someone on a background check or sexist discrimination; Find out what the fire department's appeal procedures are and decide from there your options. I would make sure you have undisputable evidence to support your concern; otherwise it's not worth the money or the time.

Final Word: Write the department or individual thanking them for the opportunity of being considered. Tell them you will not give up and that you plan to improve and overcome the reasons why you were unsuccessful this time around.

Figure 3-2
Keep track of all your recruitment information

RECRUITMENT MANAGER

City Hiring	Date	Application Return Deadline	Completed Application Package	Notes

Figure 3-3
Keep track of Fire Department recruitments information

FIRE DEPARTMENT RECRUITMENT TRACKER

Fire Department	Projected Recruitment Date mm/yr	Actual Recruitment Date mm/dd/yr	Last Recruitment Date mm/dd/yr	Comments

CHAPTER ELEVEN

APPLICATION PACKAGES

Once a fire department announces they will be recruiting probationary fire-fighters your first step is to pick-up an application form. Sometimes information regarding application forms or when they are available is not clearly stated. Immediately contact the fire department or the city's human resources department by email, phone or locate the 'recruitment announcement' ad in the local paper for further information. Fire departments use application forms as a way to standardize and outline every worthy applicant's qualifications. Your job is to make sure you complete the application in a neat manner and that the information is accurate. Make a rough draft to practice writing answers in pencil then use the original to neatly and carefully write the answers.

A city's recruitment process can be quite extensive. There can be hundreds of applications to review. Each city sets a minimum standards or a marking system that helps in narrowing the numbers down to a manageable amount. This could be good news or bad news for you depending on the depth of your résumé. When a fire department announces a recruitment make sure you get all the vital available information about the hiring. Be sure to check the fire department's website because there may be important information about the recruitment you may need to know. For example, the Calgary fire department may only be accepting application packages from candidates who reside in Calgary. So make sure you're aware of all there is to know. Other key information you want to be clear about is dates, deadlines, qualifications requirements, whether application forms are mailed our or can be picked up, and if there is a fee to pick up an application package. If you're not sure about something take the time to call to get clarification from either the fire department or city human resources.

It's important when assembling your application package that you put your documents in the proper order. Fire departments don't have the time to search for the information they want. It's vital that you assemble your application package properly.

Unless otherwise stated, the order of your application package should be assembled as follows:

1. Fire department application form (completed and signed)
2. Cover letter
3. Résumé
4. Prerequisite certificate, licences and diplomas
   Note: Colour photocopy of course certificates, licences and diplomas (place in the same order as the application form requires)
5. Fire related certificates, diplomas, degrees and transcripts
6. Non-fire related certificates, diplomas, degrees and transcripts
7. Optional Receipts for incomplete course
8. Optional copies of previous tests results
9. Military service record
10. Awards—(academic, work related, personal achievements, etc.)
11. Criminal record search documents
12. Drivers abstract/licence
13. Volunteer documents (location, hours served, duties, etc)
14. Reference Letters
15. Method of Payment (cash, certified cheque, money order, debit, etc.)

IAFF MEMBER WINS OLYMPIC GOLD MEDAL IN TURIN

FEBRUARY 17, 2006

Duff Gibson, Calgary, Alberta fire fighter and member of IAFF Local 255, edged a fellow Canadian to win the gold medal in the Men's Skeleton event during competition on Feb. 17, setting a track record in the process.

Mark Faires, President of IAFF Local 255 Calgary, said his members are jubilant over Gibson's win.

"We've been rooting for him for many years. Duff's a well-respected professional among his peers in Calgary and he's obviously a dedicated athlete," Faires said, noting Gibson's physical abilities are amazing.

"He's a pretty impressive guy when you see him working out, and he's a good fire fighter, so we're really excited for him. We're bouncing off the walls here."

Gibson, 39 works out of Calgary's No. 27 Fire hall, which serves the city's airport. He only took up the sport at age 32, but won the world title in 2004.

Faires said despite the heavy training and competition schedule, Gibson always puts his job as a fire fighter first. The entire local, and the Calgary Fire Department, have been extremely supportive of him as he pursued his dreams. Fellow fire fighters, for example, covered shifts for him to accommodate his schedule.

The skeleton sport involves a thrilling face-first slide down an icy track at speeds of up to 120 kilometres per hour. The object is to keep as straight a path as possible through the course's twists and turns.

In a 2005 interview with the IAFF Canadian Journal, Gibson discussed his growing reputation in the sport, and pondered his chances at the Olympics. "It's a technical track with 19 turns, which is on the high side. Steering around turns is one of my strengths, so I've got an edge over some of my competition."

www.iaff.org

APPLICATION PACKAGE CHECKLIST

APPLICATION PACKAGE INCLUDES:

	YES	NO
Application Form	☐	☐
Cover Letter: Up-dated résumé as of _____ (Date)	☐	☐
Prerequisite certificate, licences and diplomas	☐	☐
Fire related certificates, diplomas, degrees and transcripts	☐	☐
Non-fire related certificates, diplomas, degrees and transcripts	☐	☐
Optional—receipts for incomplete course	☐	☐
Copies of previous tests results (Optional) Aptitude Test	☐	☐
Military service records and decorations	☐	☐
Awards—academic, work related, personal achievements, etc.	☐	☐
Criminal record search documents	☐	☐
Drivers abstract/licence	☐	☐
Volunteer documents (location, hours served, duties, etc)	☐	☐
Reference Letter(s) (Only if requested)	☐	☐
Method of Payment (cash, certified cheque, money order, debit, etc.)	☐	☐

QUESTIONS TO ASK YOURSELF

	YES	NO
Is it organized?	☐	☐
Is all my information accurate?	☐	☐
Any up-dates needed?	☐	☐
Are all certificates color photocopied?	☐	☐
Should I submit my application by registered mail?	☐	☐
Is my name clearly marked?	☐	☐
Do I have the correct mailing address?	☐	☐
Is my cover letter personally addressed?	☐	☐
Is my application package easy to read?	☐	☐
Is my data in the proper order?	☐	☐
Do I possess the minimum requirements?	☐	☐
Is my contact information visible?	☐	☐
Is it in good condition?	☐	☐

CHAPTER TWELVE

HOW TO SUCCESSFULLY COMPLETE AN APPLICATION

It will seem redundant filling out an application form because you'll find that you are literally copying your résumé onto another piece of paper. But the advantage of an application form to a fire department is they can choose the order in which the information is received as well as have the ability to compare applicant's qualifications quite easily since the layout is standardized for every application.

Fire departments can easily pick out the applicants who will advance by quickly glancing at the application form and seeing if all their requested information is filled out. You could have a very impressive résumé with all kinds of courses, but if you have a lot of empty sections on your application form then you have a greater chance of being passed over.

Fire departments that require the application form to be delivered in person usually accept delivery by any person, not necessarily the applicant.

Application Form Tips

- Quickly review the entire application form before attempting to fill it out (Do not make a mark on it). This will familiarize you with the content and structure of the application form as well as help you determine what information goes where.

- Always photocopy your application form before you attempt to fill it out. Use this copy as a work sheet and keep the original as the final, finished form. This way if you make a mistake you can simply use your back-up copy—after you photocopy it.

- Study and follow all instructions very carefully. Application forms vary from fire department to fire department—and all have specific spaces in which you're expected to answer the questions.

- Complete application form as neatly as possible. Take your time when filling out your application form and start early! Don't wait until the night before you have to hand it in to start. If your handwriting is hard to read then your application may be tossed in the garbage, remember your application is a reflection of you. If available use a typewriter to complete the application.

- Always use either a black or blue pen to fill out your application form. When you make a mistake not using an erasable pen, white-out is acceptable.

- Never fold or bend your application form. It's critical you keep it in mint condition in a large envelope with all your back up data. If you need to mail your application use a waterproof and rigid letter mailer.

- Customize all category answers to be specific to the position of a probationary firefighter. Outline skills and descriptions of your experiences and achievements without including unimportant information. Follow the same structure as your résumé.

- Don't leave any box blank. Fire departments structure the application form so they get the same information from every applicant. If you come across a section on your application form that is irrelevant to you then respond with "not applicable" or "n/a". Never write "see résumé".

- Only include positive information. Never show information that gives them a reason not to give you an interview. If you provide such information like you were fired from your last job you're not sending the fire department the right message.

- Honesty is the best policy. Being caught supplying false or misleading information will terminate any potential you may have had of getting an interview.

- Application must be consistent with your résumé. Pay close attention to detail. You want to make sure that all your dates, names, titles, etc. on your application form correspond with the information on your résumé.

- Always verify the information you supply in your application. Don't just estimate or guess what years you worked for an employer. Take the time to look the information up so you can be certain it is accurate.

- Thoroughly review your application before submitting it. After you have completed your application form let it sit for a day or two and come back to it. Chances are you'll pickup on mistakes such as typos or wording that could be improved.

FIREFIGHTER APPLICATION

PERSONAL INFORMATION

Name:	
Address:	
City:	Postal Code:

Height (without shoes):	Weight:	Do you smoke? (✔) Y ☐ Amount? ___ N ☐
Visual Aids Y ☐ Required? (✔) N ☐	If Yes, describe (i.e. contacts/glasses):	
	Visual acuity without aids (i.e. 20/20. Please specify):	

Any medical conditions or disabilities that would impair your ability to perform all aspects of the job applied for: Yes ☐ No ☐

If yes, please describe your limitations:

General Health:	Social Insurance Number:		
Driver's Licence No.:	Licence Class	Air Brake (✔)	Y ☐
# of Points:	Expiry Date:	Endorsement	N ☐

Are you a Canadian citizen? (✔)If no, are you a landed immigrant? (✔)

Y ☐ N ☐   Y ☐ N ☐

EMPLOYMENT INFORMATION (list most current job first)

1.	Company Name:		Type of Business:	
	Address:			
	Supervisor's name:		Phone #:	Dates employed:
	Job Title:			Rate of pay:
	Description of duties:			
	Reason for leaving:			
2.	Company Name:		Type of Business:	
	Address:			
	Supervisor's name:		Phone #:	Dates employed:
	Job Title:			Rate of pay:
	Description of duties:			
	Reason for leaving:			
3.	Company Name:		Type of Business:	
	Address:			
	Supervisor's name:		Phone #:	Dates employed:
	Job Title:			Rate of pay:
	Description of duties:			
	Reason for leaving:			

EDUCATION AND TRAINING

	School (name/location)	Years Attended (dates)	Highest Grade Completed	Major Subjects	Did you graduate? Y/N
High School					
University or College					
Vocational or Trade Business School					
First Aid Training:					
Expiry Date:					
Can you swim? (✔) Y ☐ N ☐		Certification (provide details/dates)			
Other related training (e.g. scuba certification)					

LEISURE AND RECREATIONAL ACTIVITIES
(specify if past or present activity and time spent on fitness/current interests)

Current leisure/physical fitness activities (describe your fitness program)

Team sports and positions played/playing:

Hobbies/Volunteer work:

REFERENCES

(give three references [not relatives] who have knowledge of your character and ability)

	Name	Address	Phone Number	Occupation	Years
1					
2					
3					

GENERAL

1.	What are your reasons for wanting to become a Firefighter?
2.	Why do you feel particularly suited to this occupation?
3.	Have you made application previously to any fire department (✔) Y ☐ N ☐ If yes, provide details and dates:
4.	Additional information important to your application:

I hereby certify that all information given in this application is true and complete. I agree that any misstatement or misrepresentations herein may cause forfeiture on my part to all rights to any employment with The District of North Vancouver. I further release to the District the authority to check records and files relevant to my applications and to contact employers with regard to work references.

Signature _____ Date _____

APPLY TO DIFFERENT FIRE DEPARTMENTS

I've told you that applicants must be realistic and understand they may not be recruited by the fire department of their choice. Chances are, if you get hired it will be with the fire department of your third or fourth choice. Firefighters, once they are hired usually stay around one or two years before trying to jump ship to another department perhaps closer to home. Applicants currently involved in this stage of the recruitment process apply to as many fire departments as possible, regardless of where they are located geographically. These applicants have the right idea because the more practice you get filling out application forms and writing tests the better you will become at completing them. These applicants may spend a lot of time and money but they are often recruited by a fire department that is usually their first or second choice. When the fire department of better choice announces recruitment they are completely prepared to fill out the application form and write the aptitude test. However, there is a strategy that will improve your chances of getting hired where you prefer to be hired.

The great thing in this world is not so much where we are, but in what direction we are moving.
— Oliver Wendell Holmes

ORGANIZATION OF FIREFIGHTER RECRUITMENT DATA

Being organized with your firefighting recruitment data is very important for a number of reasons. It allows you to remain focused on what you're doing or what you need to be doing, not worrying about missing deadlines for fire department recruitment packages or the time of your interview. Can you imagine if you miss an interview because you just weren't organized? It's a must to get control of this situation at the earliest possible moment, stay enthusiastic and reduce potential stress at the same time.

To remain organized you will need some type of system in place to avoid missing dates and misplacing vital information. Localize everything you have to do with "fire." Have one area that is dedicated for "fire" information only. When you misplace anything know immediately where to start looking. Even though you have a location for all your fire recruitment data you still need to be able to locate information at a glance. The best way to do this is using a binder system one for each topic. For example, have a red binder for fire departments. Divide this binder into sections. In each section write in a fire department's name, from then on any information you receive regarding that fire department will go in its appropriate section. You will be surprised how much information you accumulate. When choosing fire departments for your binder only include the ones you are actively involved in or interested in pursuing. It is unrealistic and unnecessary to list every fire department, be selective.

For each fire department listed, use a copy of our Figure 3-4. Keeping track and updating this vital information will greatly reduce the possibilities of misplacing or losing information that you have collected.

Figure 3-4

Photocopy and complete Fire Department Profile for
every fire dept. you visit or are interested in.

FIRE DEPARTMENT PROFILER

Fire Department: _____

Phone #: _____

Website Address: _____

Address: _____

_____

_____

Chief's Name: _____

Deputy Chief's Name: _____

Human Resource Contact: _____

Other Contacts (fire halls, etc.) _____

_____

_____

Human Resources Address: _____

_____

Phone No.: _____

Fax: _____

City Website: _____

_____

_____

City Population: _____

Number of Fire Halls: _____

Fire Hall Location(s) _____ @ _____

_____ @ _____

_____ @ _____

_____ @ _____

_____ @ _____

_____ @ _____

Type of Shifts Worked: _____

Number of Emergency Calls Annually: _____

Division of Jurisdiction Regarding Fire Districts: _____

Future Growth (new fire hall, new trucks etc,) _____

_____

_____

Specialized Teams: _____

_____

_____

_____

_____

_____

_____

This will help avoid the situation of mailing the same fire department twice or even worse forgetting to mail them at all. Remember to update this information constantly.

Monitoring each fire department's website on the same day and time each week, this will ensure that you stay informed about current vital information.

FIREFIGHTERS MAKE DRAMATIC ROOF RESCUE OF WOMAN TRAPPED BY FIRE

MARCH 19, 2006

Firefighters made a dramatic rescue from the top floor of a burning building in Crown Heights Wednesday morning in a scene that looked like it was straight out of a movie.

Fire and heavy smoke spread to the upper floors of the Ebbets Field Houses in Crown Heights after flames were sparked in a trash compactor in the basement.

The fire forced one woman onto the ledge from the top floor of her apartment. When firefighters arrived at the scene, they found Cheryl Ann John dangling from the roof of the 25-story building, held only by the weakening grip of her husband and a Good Samaritan neighbor.

Three firefighters from Ladder Company 113 and Ladder 132—Joe Donatelli, Bill Hansen and Tim Rail—were able to rappel down from the roof to pull Johnson to safety.

"She ended up letting go of one hand and she put it around my neck, and I don't know if her other hand actually made it around me, but she held on to me, I held on to her, then they pulled me up," said Firefighter Donatelli of Ladder 132.

"This is the life belt that Billy put on," said Firefighter Rail of Ladder 113, holding up a safety belt that made the rescue possible. "He was the anchor. It ties around here [showing metal clip]. This goes over the parapet [holding up belt]. Joe put it on. This is the rest of the rope."

Two people, including the woman who was rescued, were taken to Kings County Hospital with minor injuries.

Three firefighters suffered minor injuries battling the blaze and three civilians, including John and her husband, and another person suffered smoke inhalation.

By: On NY1 Now

Figure 3-5

When you receive notification of a fire department hiring, complete this form.

FIRE DEPARTMENT HIRING CHECKLIST

Fire Department: _____

Phone No: _____
_____

Website Address: _____

Address: _____
_____
_____

Possible Hire Date: _____
_____

Hand-in application by: _____
_____

Cost: _____
_____

Prerequisites: _____   _____
_____   _____
_____   _____

You should also organize the following:

Fire School: If you're planning to attend a post-secondary fire college, a three ring binder must contain all the information you receive regarding the college or program, as well as any additional information such as flight information, accommodations, financing, contact information, etc.

Licences and Courses: Should have its own three ring binder in order to keep track of licences and courses you have obtained as well as for licences and courses you wish to obtain.

When completing a licence or a course it's important that you keep information about these qualifications in this binder. By having proper documentation of each licence or course you will be able to resort back to it and refresh yourself on the qualifications you have taken. Let's face it, some of the courses you've taken may be a few years old and you don't really remember the topics covered without reviewing the course content or outline. Are they out of date and need updating?

Résumé Packages: A fourth binder is necessary for Résumé Packages. You should always have an updated résumé on hand. You never know when you're going to need one. You should keep extra photocopies of your courses and licences, preferably in color on hand. Keep copies of your high school diploma, references, template cover letters, and thank-you letters on hand too.

If you want you can take this one step further, create an up-to-date preassembled résumé package. When a fire department posts a hiring, you are prepared. In this case all you have to do is insert the department's name into the already-made cover letter template and envelope sticker and you are all set. This saves a lot of scrambling and unnecessary worrying at the time of hiring. The less you have to worry about the better off you are. But it's always smart to check it over before submitting.

If what you're working for really matters, you'll give it all you've got.
—Nido Qubein

❗ KEY POINTS TO REMEMBER

- Fire Departments can announce they are recruiting at any time.
- If you are not successful in any stage of the recruitment process chances are there is a way you can overcome it.
- Always photocopy your fire departments application form, you will have a spare if you make any mistakes.
- Be sure to include all necessary documents with your application package.

NOTES

_____

_____

_____

_____

_____

_____

_____

_____

_____

NOTES

Part IV

Test Strategies:
Written & Physical

CHAPTER THIRTEEN

PASSING THE APTITUDE TEST

The entry firefighter tests are designed to help fire departments assess the qualifications of each applicant. It is specific to entry-level firefighter applicants with no previous fire related education or experience. It measures applicant's abilities in a number of areas including their understanding of written and oral information, mechanical aptitude, mathematics, as well as the candidate's ability to handle people. The multiple choice format has no penalty deductions for wrong answers.

The number of test writing assemblies is determined by the number of applicants involved. Usually a fire department will choose one or two days for aptitude test writing. It is not uncommon for a fire department to use a large assembly hall as they have hundreds of applicants writing at the same time. On the day of the test be sure to arrive early. Before you are allowed to write the test you have to check in with the testing authority. Fire departments will have a list of candidates who are eligible to write the aptitude tests. Often the candidates will have to provide some photo identification as well as pay the test-writing fee. Once verified, you then proceed to the test writing room. Once inside you will take a seat and wait for the test to begin. Instructions will be provided on how to properly fill in the information such as your

Unsuccessful applicants will be returned to the Applicants Pool for further consideration if other positions become available.

name, test number, the date and time, location of the test and so on. After this vital information is completed you will begin the test.

SCORING OF TESTS

Scores on the tests are intended to indicate knowledge of the subject matter emphasized in the fire service. Because past achievement is usually a good indicator of future performance, the scores are helpful in predicting who will be rated as a quality firefighter applicant. The tests are standardized throughout each fire department's recruitment process, and test scores allow comparison of candidates with different experiences and education. In order to pass the test candidates must attain a mark equal to or higher than a pre-determined threshold mark, which is usually 70% or higher.

In order to give applicants a better breakdown of their marks some fire departments use sub-scores. The sum of all the sub-scores equals the total score. Notice how "Reading Comprehension" provides the greatest amount of points to the total of 100.

SUB-SCORE	
Oral passage –	16/20
Reading comprehension –	18/25
Mathematical –	16/20
Mechanical aptitude –	14/20
Dealing with people –	14/15
TOTAL SCORE –	**78/100**

The written aptitude test has passing scores that are predetermined and not based on the number of applicants.

Most fire departments use the scores on the entry firefighter test as a general candidate qualifier, which is considered with other relevant information about every applicant. Because numerous factors influence success in the fire service, reliance on a single measure to predict success is not reasonable. Other indicators of competence typically include a résumé package outlining courses obtained and grades earned from any post secondary education, references and the physical assessment scores.

Every potential firefighter is expected to successfully pass this portion of the recruitment process in order to advance and be considered for an interview. Most if not every fire department uses a written aptitude test. Chances are you will be required to write an entry level firefighting test at some point throughout your recruitment process. Some candidates may apply to several fire departments and find themselves writing several exams. There is a possibility that different fire departments use the same entry firefighter exam, but for the most part the tests with be different. Most fire departments contract the responsibilities of creating entry firefighter tests to independent test writing agencies. This saves both staff time and money when compared to having the fire department create their own specific entry firefighter tests. However, some fire departments continue to create their own tests and have had great success in doing so.

CHAPTER FOURTEEN

CPS FIREFIGHTER TEST

The CPS (Cooperative Personal Services) test consists of 100 questions and is made up of five different categories.

- **Listen-to Passage** (20 Questions)
  The scenario will be read to the group in under 10 minutes or there may be two smaller sections read back-to-back ranging from 1 1/2–3 minutes.
- **Mathematics** (20 Questions)
- **Mechanical Aptitude** (20 Questions)
- **Reading Comprehension** (25 Questions)
- **Dealing with People** (15 Questions)

The CPS test is the most common entry firefighting test you will find in Ontario. The rest of Canada and the United States use tests that resemble the CPS tests. Categories commonly found on these tests include any number of the following (not in any particular order).

- Memory of understanding oral information
- Mapping

- Understanding written firefighter material
- Grammar
- Mathematical reasoning
- Understanding three-dimensional diagrams
- Mechanical aptitude
- Construction
- Tool recognition
- Teamwork
- Community living

Each category is then given values that help fire departments grade each test. The test you write could have all of these topics or only a few. It really depends on what categories are important to the fire department that created the test. It's important to be prepared as well as confident when entering a firefighting test. You don't have to be an experienced firefighter to do well. With a grade 12 education or equivalent, you should have no problems completing the mathematical portion on this exam.

CHAPTER FIFTEEN

PREPARING FOR WRITTEN APTITUDE TESTS

To help repress test anxiety I recommended writing as many entry-level firefighter tests in order to become more comfortable in writing them; "Perfect practice makes perfect". Don't wait to write entry-level firefighter tests. How would you feel if the fire department you've been waiting for posts a recruitment and you write the exam and fail? If you don't want this to happen to you, prepare yourself. Practice writing exams and answering questionnaires.

Here are some key points when preparing for your test:

TIP!

There are also firefighter exam books you can buy that have questions very similar to actual firefighting exams.

www.chapters.indigo.ca

- Only highlight the areas in the questions you have problems with. Continue to go over these areas until you understand the strategies and philosophies behind them. It is better to find out before you write the test that you don't understand the question than during. This will allow you time to go over those areas in more detail.

- It is important to free yourself from distractions such as radios, television or even people talking while studying.

- Adequate lighting helps increase your concentration when studying.
- If you smoke you may find it difficult to last 2 to 3 hours without having a cigarette. You should practice not smoking for 2 to 3 hour periods, so you don't have to worry about dealing with anxiety or frustration from not having a smoke during the examination. (You should quit)
- Candidates who are physically fit will also have a fit mind. Be sure to stay in shape and maintain your physical fitness level and stay away from stressful situations to keep your mind clean at examination time.
- Rest is extremely important not only the night before the test but the week prior to the test. Try to stay consistent with your regular sleeping habits. For example if you are accustomed to getting 6 hours sleep a night be sure to get a at least 6 hours sleep the night before the test. I would also not recommend receiving 11 hours sleep the night before because chances are you will probably wake up feeling groggy for the rest of the day.

Figure 4–1
Steps taken when answering problem solving questions.

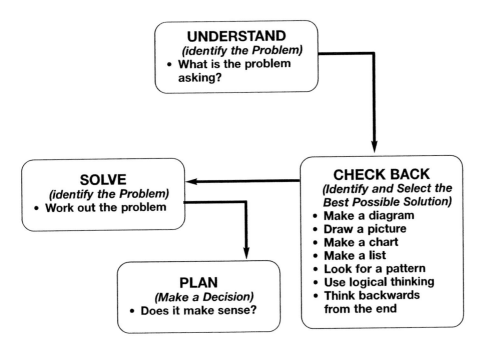

Eliminate Choices That:

- Benefit you alone
- Require that an order be obeyed
- Call for unnecessary actions that are contradictory to what is required
- Only partially solve the problem
- Insult, disregard, interfere with, or hazard a citizen

- Make sure the day of your exam you leave early and allow extra time for traffic problems and complications. Be in the examination room 20 to 30 minutes ahead of time: You will require this extra time to become accustomed to the environment of the room. If you're too hot or too cold, or need a bathroom break you can address the problem in time to write your exam. Remember to layer your clothing; it's easiest to remove extra clothing if you are too warm.

- Once in the examination room you will be given forms to fill out. The examiner will give specific instructions such as how to fill in the grids on the forms, what the time limits are and the timing signals to be given. If at any time you do not understand any of these instructions raise your hand to ask a question. Don't cut in when someone from the department is speaking. Make sure you understand exactly what to do, and how long you will have to do it. Stay calm.

ENTRY-LEVEL FIREFIGHTER TEST WRITING TIPS

Listening Skills

Sit near the front to eliminate distractions and help your focus and hearing. The listening portion of the test has the potential to be the most difficult due to several factors. One factor being it's the first part of the exam and chances are your nerves are trying to get the best of you. Another reason is simply the fact that you spend so much time trying to pick out specific details that you miss the whole picture. There are things you can do to improve your final results:

1. Eye contact: It's easier to remain focused during the reading, or speaking if you maintain constant eye contact with the reader. This will help avoid being distracted by what's around you as well as prevent day dreaming while the passage is being read to you. You can't afford to miss any information.

2. Concentrate on the content, not the delivery. It's easy to be distracted if you find the reader amusing because of the way he/she speaks, the hand gestures that he/she makes or the number of times he/she blinks while reading. Stop noticing these things and focus only on the content of their delivery. Get your imagination and memory in high gear.
3. Don't get too emotionally involved. If you're too emotionally involved when listening, you may make the mistake of hearing what you want to hear and not what is actually said. Remember not to be biased and opinionated. Keep an open mind.
4. Don't be a victim of distractions, sit near the front. Maintain your focus on the reader when the person next to you is tapping their pencil on the table.
5. Controlling body temperature. Remove layers if you are too hot.
6. Control the time between the rate of thought and the rate of speech. Did you ever wonder why you lose your concentration so easily? It's because your mind can think faster than someone can speak.

To avoid becoming bored, try to anticipate the end of the sentence. It's possible to listen, think, write, and ponder all simultaneously. It just takes practice.

Reading Comprehension

The reading comprehension section of the test measures both literacy and your understanding of text. You will be provided with a short reading passage and followed by several multiple choice questions to test your comprehension. All the answers to the questions can be deduced from the passage itself and usually consist of answering technical information, not resembling anything like a novel study or essay. Time is very important when writing these tests to help you to concentrate and read faster. Use your pencil to follow the words as you read along. This helps you focus on the details in the reading (easy to underline something to remember) and increases your concentration on the task at hand.

Take the time to circle or underline the key words and phrases that you feel are important in the passage. This will allow you to focus on only the vital information required to answer the questions. For short passages 6 to 7 sentences in length take the time to absorb the passage carefully. You should have the ability to retain all of the essential information in the reading. By making the extra time and effort to read the passage carefully, you only have to read it once. Before attempting to read longer passages take a moment and peruse the questions you are being asked. Sometimes reading passages are lengthy, but the questions are only asking for specific details. Your memory works better if you can picture what you read—or, better yet, place yourself in the passage. For example, if you drive by a car accident you may say, "I hope no one is injured" but if you were in the car accident you will remember it for the rest of your life. By placing yourself in the reading you will aid the placement of the information into your long-term memory. Then you will be able to recall the information as you need it to answer the questions.

Don't be intimidated by the length of the passage or the information in it. These tests were not written to intimidate or discourage you. They are simply testing your reading comprehension and your ability to recall information from what you have read. You can increase your chances of doing well in this section by understanding how you handle questions like this. For example you may find it easier and quicker not to read the passage and just look for the information they are asking for in the questions, or if you have to read the entire passage to be able to answer each question. It is important to understand yourself and how you work. Everyone works and thinks differently. If you're not sure how you would handle a question like this, you can practice by having a friend read a newspaper article and write questions about the content. Then they can give you the article to read and have you answer the questions. The more you practice answering this type of question, the more relaxed and confident you will become in completing these types of questions.

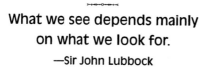

**What we see depends mainly
on what we look for.**
—Sir John Lubbock

Figure 4–2

Read scenario and answer the questions that follow.

SAMPLE READING COMPREHENSION QUESTIONS

Scenario:

Fire extinguishers are probably the most common fire-protection appliance used today. You can find fire extinguishers in most homes, commercial businesses, industrial work places, and commercial vehicles. Types of fire extinguishers may vary but they all share the same principal, fire extinguishment. Fire extinguishers are only effective for extinguishing incipient fires, often a fire extinguisher can extinguish a small fire in less time than it would take to deploy a hose line to put out the fire. There are regulations, such as NFPA 10 (standard for portable fire extinguishers) that regulates the rating, selection, and the inspection of portable fire extinguishers.

There are different types of portable fire extinguishers: some common ones are pump tank water, stored pressure water, halon, carbon dioxide, dry chemical and dry powder. All are effective if used on the proper classification of fire. Without selecting the proper fire extinguisher you may cause more damage than good.

The four classifications of fire are class "A" fires, ordinary combustibles such as paper and wood. A class "B" fire is flammable liquids such as gasoline or oil. The third class of fires is a class "C" fire used for energized electrical fires (if power is de-energized the class "C" fire turns into a class "A" fire). The fourth classification of fire is class "D" fire, which are combustible metals such as magnesium, and/or sodium. Portable fire extinguishers that are suitable for more than one class of fire are identified using different combinations of letters A, B or C . The most common types of combinations are class A, B, C, class A-B, and class B-C. Portable fire extinguishers must be properly marked or the extinguisher should not be used. Along with the class of fire the extinguisher will be used on it should also have the amount of extinguishing agent? For example, if the fire extinguisher may have a rating of 4-A 20 B:C. The extinguisher should extinguish a class A fire and extinguish 4 times larger than a 1-A fire, extinguish 20 times as much class B fire as a 1-B extinguisher, and extinguish a deep- layer of a 20 square- foot (2 m squared) area. This extinguisher can also be used on fires involving energized electrical equipment.

These fire extinguisher ratings are for untrained users. A trained individual would be able to put out a greater amount of fire. To properly select a portable fire extinguisher, there are things you need to consider such as classification of the burning fuel, rating of the extinguisher, hazards to be protected, severity of the fire, atmospheric conditions, availability of trained personnel, ease of handling extinguisher, and any possible life hazards or operation concerns.

When using an extinguisher remember the acronym PASS P-pull the pin, A-aim, S-squeeze the trigger, and S-sweep back and forth over the fire. Using an extinguisher can prevent a little accidental grease fire in a frying pan on a stove to becoming a house burning to the ground. Make sure that if you have an extinguisher, you know its limitations and become familiar with how to use it.

1. Where else can portable fire extinguishers be found besides homes, commercial businesses and industrial work places?
   A) Domestic vehicles
   B) Commercial Vehicles
   C) Motor Homes
   D) Boats

2. Fire extinguishers are effective in extinguishing what types of fires
   A) Garbage fires
   B) Fully developed fires
   C) Incipient fire
   D) Free burning

3. Fire extinguishers can put out small fires:
   A) In the same time it takes to deploy a hose line
   B) In less time than it takes to deploy a hose line
   C) In more time than it takes to deploy a hose line
   D) Only if the fire is outdoors

4. Regulation for portable fire extinguishers is?
   A) NFPA 12
   B) NFPA 10
   C) NFPA 15
   D) NFPA 20

5. What are the four classifications of fire?
   A) 1, 2, 3, and 4
   B) Orange, yellow, black, and green
   C) A, B, C, and D
   D) None of the above

6. What are class "A" fires?
   A) Ordinary combustibles
   B) Energized electrical
   C) Flammable liquids
   D) Combustible metals

7. What is a class "C" fire?
   A) Ordinary combustibles
   B) Energized electrical
   C) Flammable liquids
   D) Combustible metals

8. What are the three combinations of fire extinguishers?
   A) Class A, class B: C, class A: B
   B) Class A: B: C, class C, class A: B
   C) Class A: B: C, class A: C, class B: C
   D) Class A: B: C, class A: B and class B: C

9. What do you need to consider, when selecting a fire extinguisher?
   A) Classification of burning fuel
   B) Severity of the fire
   C) Both A & B
   D) None of the above

10. What does PASS stand for?
    A) Pull pin, aim, safety glasses and stance
    B) Pull pin, aim, squeeze trigger, and sweep
    C) Pick the extinguisher, aim, squeeze, and sweep
    D) Pull pin, aim, squeeze, and safety

www.becomingafirefighter.com

Use the sample memory tests provided on the website to practice your reading comprehension skills.

Answers:
1. B; 2. C; 3. B; 4. B; 5. C; 6. A; 7. B; 8. D; 9. C; 10. B

CHAPTER SIXTEEN

STRATEGIES TO ANSWER MULTIPLE CHOICE TESTS

When you take the test, you will mark your answers on a separate machine-scored answer sheet. Total testing time is two hours; there are no sections separately timed. The following are some general test-taking strategies you may want to consider.

- Read over the entire test before attempting to write it. Make sure you are not missing any pages and that you are writing the right test.
- Read the test directions carefully, and work as rapidly as you can without being careless. For each question, choose the best answer from the available options. Use an erasable pen.
- If you decide to change an answer, make sure you completely erase it and fill in the spot corresponding to your desired answer. Most tests are graded by computer and the computers aren't designed to decipher your respective answers.
- Make sure you monitor the time periodically throughout your exam. For example, if the test has 200 questions and you are allowed 2 hours to write the exam, then 1 hour into the exam you should be roughly on question 100.

- Never wait until the last few minutes to record your answers.
- Eliminate options you know to be incorrect. If allowed, mark words or alternatives in questions that eliminate the option. Give each option of a question the "true-false test": This can help to reduce your selection to the best answer.
- You may want to work through the test, by first answering all the questions that are easy or about which you feel confident, then start again to answer questions that require more thought, and concluding with the most difficult until time runs out.
- "All of the above": If you know two of three options seem correct, "all of the above" is a strong possibility to use as an answer.
- Questions for which you make no answer or give more than one answer are not counted in scoring.
- Number answers: Toss out the high and low and consider the middle range numbers.
- Double negatives: Make it into the equivalent positive statement and consider.
- Echo options: If two options are opposite each other, chances are one of them is correct.
- Favor options that contain qualifiers. The result is longer, more inclusive items that better fill the role of the answer.
- Always guess when there is no penalty for guessing or you can eliminate options.
- Don't guess if you are penalized for guessing and if you have no basis for your choice.
- Use hints from questions you know to answer questions you do not.
- Remember that you are looking for the best answer and not only a correct one—and not one which must be true all of the time, in all cases, and without exception.

During multiple choice tests it can be difficult to pick the correct answer. Just eliminate the wrong ones that are obvious.

Studies have proven that when answers are changed on tests 75-80% of the time answers have been changed from the right answers to the wrong answers.

Do not memorize the questions and answers for firefighter test. Make sure you understand how the answers are derived from in each type of question. You don't want to rely on the possibilities that the questions are the ones you have memorized.

HOW TO FIND THE CORRECT ANSWER

Selecting the correct answers on a multiple choice test can be difficult at times. Usually you can eliminate two of the four answers right away. The tricky part is now deciding which one out of the two answers remaining is the correct one. Paying close attention to the wording of the answers could give you hints which answers are more than likely to be correct or incorrect. If the answer is limited or overstated its probably incorrect. If the answers are exact or vague, it's probably correct.

Example:

A firefighter is required to wear a pass (Personal alarm safety system) alarm at any fire scene where the firefighter is required to enter any structure wearing SCBA. If a firefighter's pass alarm fails to work properly, the officer in charge will be notified of the problem immediately. Chances are the firefighter will be reassigned to a low risk task such as managing the accountability board until the situation has been brought under control or the firefighter attempts to get the pass alarm working properly and is reassigned back to the structure.

Based on the paragraph above, select the correct answer.

A) A firefighter should wear the pass alarm only when entering a structure wearing SCBA

B) A firefighter must attempt to repair a pass alarm which is not working properly

C) A firefighter whose pass alarm is not working properly will be assigned to the accountability board

D) The firefighter whose pass alarm has been repaired can be reassigned back to the structure.

Certain words in the above answers can help determine whether the answer is correct or incorrect.

- The A) option is incorrect because it contains the word "only".

  A) a firefighter should wear the pass alarm only when entering a structure wearing SCBA

  Note: *Without the word "only" A) would be the correct answer.*

- The B) option would be correct if it said "may" instead of "must".

  B) a firefighter must attempt to repair a pass alarm which is not working properly

- The C) option is incorrect because it says "will".

  C) a firefighter whose pass alarm is not working properly will be assigned to the accountability board

  Note: *C) would be better if it said "may" instead of "will" answer.*

- The D) option is correct; it only says "can".

  D) the firefighter whose pass alarm has been repaired can be reassigned back to the structure.

  Note: *D) would be incorrect if it said "must" instead of "can".*

When selecting your answer be sure to watch for words that limit or overstated such as the following:

• no one	• invariable
• all	• surely
• every	• any
• certainly	• no matter
• always	• nothing
• will	• ever

Words which are less exact and/or restrictive such as the following make the answer more appealing:

- many
- sometimes
- may
- some
- possibly
- can
- often

- might
- usually
- could
- occasionally
- generally
- probably

THE LITTLEST FIREFIGHTER

In Phoenix, Arizona, a 26-year-old mother stared down at her 6 year old son, who was dying of terminal leukemia. Although her heart was filled with sadness, she also had a strong feeling of determination. Like any parent, she wanted her son to grow up and fulfill all his dreams; now that was no longer possible. The leukemia would see to that. But she still wanted her son's dreams to come true. She took her son's hand and asked, "Billy, did you ever think about what you wanted to be once you grew up? Did you ever dream and wish what you would do with your life?"

"Mommy, I always wanted to be a fireman when I grew up."

Mom smiled back and said, "Let's see if we can make your wish come true." Later that day she went to her local fire department in Phoenix, Arizona, where she met Fireman Bob, who had a heart as big as Phoenix. She explained her son's final wish and asked if it might be possible to give her six-year-old son a ride around the block on a fire engine.

Fireman Bob said, "Look, we can do better than that. If you'll have your son ready at seven o'clock Wednesday morning, we'll make him an honorary fireman for the whole day. He can come down to the fire station, eat with us, go out on all the fire calls, the whole nine yards! And if you'll give us his sizes, we'll get a real fire uniform for him, with a real fire hat-not a toy one-with the emblem of the Phoenix Fire Department on it, a yellow slicker like we wear and rubber boots. They're all manufactured right here in Phoenix, so we can get them fast."

Three days later Fireman dressed him in his fire uniform and escorted him from his hospital bed to the waiting hook and ladder truck. Billy got to sit on the back of the truck and help steer it back to the fire station. He was in heaven. There were three fire calls in Phoenix that day and Billy got to go out on all three calls. He rode in the different fire engines, the paramedic's van, and even the fire chief's car. He was also videotaped for the local news program. Having his dream come true, with all the love and attention that was lavished upon him, so deeply touched Billy that he lived three months longer than any doctor thought possible.

One night all of his vital signs began to drop dramatically and the head nurse, who believed in the hospice concept that no one should die alone, began to call the family members to the hospital. Then she remembered the day Billy had spent as a fireman, so she called the Fire Chief and asked if it would be possible to send a fireman in uniform to the hospital to be with Billy as he made his transition.

The chief replied, "We can do better than that. We'll be there in five minutes. Will you please do me a favor? When you hear the sirens screaming and see the lights flashing, will you announce over the PA system that there is not a fire? It's just the fire department coming to see one of its finest members one more time. "And will you open the window to his room?"

About five minutes later a hook and ladder truck arrived at the hospital and extended its ladder up to Billy's third floor open window 16 firefighters climbed up the ladder into Billy's room. With his mother's permission, they hugged him and held him and told him how much they loved him.

With his dying breath, Billy looked up at the fire chief and said,

"Chief, am I really a fireman now?"

"Billy, you are, and the Head Chief, Jesus, is holding your hand," the chief said.

With those words, Billy smiled and said, "I know, He's been holding my hand all day, and the angels have been singing..." Then he closed his eyes one last time.

By Julia Meadows, *Landmarks Magazine*

CHAPTER SEVENTEEN

PHYSICAL FITNESS TEST

Firefighting is one of the most physically demanding professions. Firefighters require high levels of fitness to perform and execute all fire ground operations and tactics. Firefighters must rely on physical fitness to maintain a tolerable body core temperature, blood pressure and heart rate when exposed to a heavily heated environment for a sustained period of time.

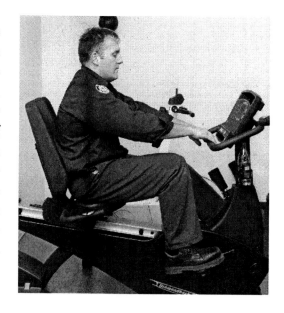

Fire departments use testing to ensure firefighter recruits have an above average fitness level in flexibility, cardiopulmonary endurance, muscular strength and muscular endurance. You must meet each of these standards before they offer you full employment. So eventually your time will come when you have to complete and pass a physical fitness test. Some fire departments will offer you a job on the condition you complete and pass a recognized physical fitness test while other fire departments like you to possess a valid physical fitness test prior to being offered employment and the documentation to prove it. These tests are usually valid for six months and cost from one hundred twenty dollars to two hundred dollars. Some candidates choose to take the expensive route and always have a valid physical fitness test handy, meaning every six months they're spending one hundred to two hundred dollars just in case they need it. I suggest you wait until you need a valid physical fitness test before you get one. In most recruitment situations you're given enough notice to make an appointment.

A couple of different types of physical fitness tests are available and although each testing facility offers a similar test, not all are recognized by the fire department hosting a recruitment. It could happen where you apply for three fire departments and have to take three different physical fitness tests. Obviously if you pass one you can pass them all but unfortunately this is the way it is until the growing pains subside. Eventually you will have to take just one physical fitness test and it will be valid for any fire department.

It doesn't matter what physical fitness test you take, the responsibility to prepare yourself for the test is yours. Some places that offer the tests have extra help available such as practice time. Make the extra effort to find which physical abilities test you will perform and exactly what tasks you will be asked to perform. To better prepare yourself, talk to candidates who have already completed the physical fitness test you have to do. Ask questions that may give you an advantage or at least tell you what to expect.

You will have enough notice to figure out what type of physical fitness tests you will be performing. You can adjust your workout to incorporate specific events of your test. If you find that you require more upper body strength, use this knowledge when preparing your workout. If you aren't sure which exercises to incorporate into your workout, explain your situation to a fitness trainer at your local gym and see what they suggest to do—and also consult the internet.

You see things; and you say "Why?'
But I dream things that never were; and say, "Why not?"
— George Bernard Shaw

CPAT—PHYSICAL FITNESS TEST

The CPAT (Candidate Physical Aptitude Test) will eventually become the standard fitness assessment test throughout Canada. CPAT uses a pass/fail method of grading each candidate; this is based on the completion of each event in concession as well as completing all events within the allowed time frame of 10 minutes and 20 seconds. CPAT is unique because each prop used has been clinically tested to insure consistency, safety as well as providing fire departments an accurate assessment of simulated firefighting tasks. Throughout the duration of the test each candidate must wear additional weights to simulate bunker gear (50 lbs or 22.8 kilogram vest) and firefighting equipment (self contained breathing apparatus). To simulate a high-rise pack the stair climb is the only event where additional 25 lbs or 11.3 kilograms is used.

Each candidate participating in the fitness test will be assigned an evaluator to provide guidance as well as time and evaluate each event. Mimicking in order the sequence of actual fire ground activities each candidate must complete each job task. Between each event the candidate is provided with an 85-foot (25.91-m) walk. Use this time (20 seconds) to recover and prepare for the next event.

CPAT consists of 8 events (in order):
1. **Stair Climb**
2. **Hose Drag**
3. **Equipment Carry**
4. **Ladder Raise and Extension**
5. **Forcible Entry**
6. **Search**
7. **Rescue**
8. **Ceiling Breach and Pull**

1. **Stair Climb**
   This event is designed to simulate climbing four flights of stairs in full protective clothing while carrying a high-rise pack.
   - Must carry an additional 25 lbs (simulate high-rise pack).
   - You will be given a 20 second warm-up period on the stepmill.
     (The rate will be approximately 50 steps/min)

- Once warm-up is completed the 3 minute test of 60 steps per minute will begin. Candidates may only use the hand rails during the test to re-establish balance.
- Walk 85 feet to the next event.

2. **Hose Drag**

This event simulates the dragging of an uncharged hoseline from the fire apparatus to the fire.
- Candidates must drag a 200 foot length of 1 3/4 inch fire hose.
- Drape a nozzle attached to 100 feet of 1 3/4 inch hose over your shoulder or across your chest and dragging in 75 feet make a 90 degree turn and continue dragging the hose an additional 25 feet to a mark indicated on the ground.
- Candidates must then stop and pull 50 feet of hose towards themselves from the kneeling position past the mark on the ground.
- Walk 85 feet to the next event.

3. **Equipment Carry**

This event simulates the retrieval of power tools from the fire apparatus, carrying them to the emergency scene and returning them to the fire apparatus.
- Candidate removes two power saws from a cabinet one at a time and sets them on the ground.
- Pick-up saws and carry them 75 feet to a marker, turn around the marker and carry them back to the cabinet.
- Place tools back in the cabinet.
- Walk 85 feet to the next event.

4. **Ladder Raise and Extension**

This event simulates the placing of a ground ladder at a structure fire and extending the ladder to the roof or a window.
- Event consists of two 24 foot extension ladders.
- Candidate must first raise a 24 foot extension ladder by placing the butt of the ladder against the wall and lifting the ladder over his/her head at the opposite end. Walk towards the wall using a hand-over-hand method, gradually raising the ladder until it is upright (vertical).

- Once first ladder is raised, the candidate must quickly take hold of the second 24 foot extension ladder and extend the fly of the ladder to reach a predetermined height and then bed (or lower) the ladder back to its original position.
- Walk 85 feet to the next event.

5. **Forcible Entry**

This event simulates forcing a locked door or the breaching of a wall.
- Candidates use a 10 pound sledge hammer to strike the side of a device mounted 39 inches off the ground.
- Buzzer indicates when you are to stop.
- Place the sledge hammer on the ground.
- Walk 85 feet to the next event.

6. **Search**

This event simulates searching for a fire victim with limited visibility in an unpredictable area.
- Start off crawling through a dark tunnel 3 feet high by 4 feet wide.
- The tunnel is 64 feet in length with two 90 degree turns.
- Candidate must advance through this tunnel navigating over and under obstacles along the way. There are two locations in the tunnel where candidates must crawl through narrow space.
- Once exiting the tunnel at the other end the event is over.
- Walk 85 feet to the next event.

7. **Rescue**

This event simulates the removal of a victim or injured firefighter from a fire.
- Candidates must drag a weighted mannequin 45 feet to a marker and then make a 180 degree turn around the marker and return 35 feet to the finish line.
- You may release the mannequin to get a better grip.
- To complete the course the mannequin must be brought completely across the finish line.
- Walk 85 feet to the next event.

8. **Ceiling Breach and Pull**

This event is designed to simulate the breaching and pulling down of a ceiling to check for fire extension.

- Remove the pike pole from bracket.
- Stand under simulated ceiling and extend the tip of the pike pole to push open a 60 lb hinged door 3 times.
- Then using the hook you must pull down an 80 lb device from the ceiling 5 times.
- You must complete 4 revolutions of the event.

Test day is not to be taken lightly; make sure you are there on time. Failure to do so could cause you to miss important instructions regarding the test or worse, disqualify you. Having water and a source of energy such as a nutritional bar with you for your test is a good idea. Quite often shower facilities are available for you to use on site. Bringing a towel and a change of clothing is recommended.

Don't have the mentality that you are in this alone. It will be a long day waiting around by yourself. Be positive, make a friend or two; it's always nice to have someone encouraging you when you're performing in an event. The camaraderie is sometimes overwhelming—that is what firefighting is all about. Chances are, not everyone taking the test that day is taking it for the first time. For some, it may be the third or fourth time, so if you have any questions, chances are someone with more experience will be able to help you.

Make certain that you fully understand every instruction given. Listen very carefully and concentrate! If you are uncertain about a detail, it is better to be centered out and show your confidence by asking a question than it is to perform the event incorrectly.

If the equipment you have chosen doesn't fit you right, get some that does. If you fail an event or a task because your equipment failed you; you will regret it. Make your first try the only one needed. You must control the things that you can. If you are prepared in mind and body, the greater will be your chance for success.

TIP!

When using ankle weights make sure they are tight around your ankles. If you become fatigued loose ankle weights will hinder your natural motion of stepping and could essentially lead to injury.

PREPARING FOR THE PHYSICAL FITNESS TEST

Listed below are some simple effective ways to prepare for your physical fitness exam.

- To decrease the chance of cramps during your test, be sure to saturate your body with water the day before the test. Drink water until you have to urinate 15 minutes after taking another drink. Additionally, you should drink at least one liter of water one hour before your physical fitness exam starts.
- To increase your level of fitness you must have an effective cardiovascular condition. You want to achieve a heart rate of around 140 to 160 beats per minute. Practice keeping this heart rate for 20 to 30 minutes by jogging or exercising.
- Take the time to warm up before every event, for the best results try to have your warm up simulate the event in which you will perform. For example if you're preparing to run you should run in place for at least two minutes or for a short distance at a very easy pace.
- Stretching for ten minutes before and after each event is recommended. Hold each stretch for 10 seconds in a range of motion that produces only mild tension. Then you should move slightly farther to the point where you feel a little more tension. Hold this for another 10 seconds.

www.becomingafirefighter.com

Become a member and take advantage of the "Workout Creator & Nutritional Menu" feature. Available exclusively to members, you can create your very own workout. Once you have designed your workout choose a nutritional menu to compliment it.

FIREFIGHTER FUND RAISES FOR NEW ORLEANS COLLEAGUES

Darryl Wiechman, a volunteer fire fighter for the RM of MacDonald (just south of Winnipeg) is selling t-shirts to fire fighters across the country to raise money for New Orleans' beleaguered fire fighters, 80 per cent of whom are still homeless as a result of last fall's Hurricane Katrina. The t-shirts feature a fire truck coming out of a circle on the front and the slogan, "Brothers Helping Brothers" on the back.

Wiechman is selling the t-shirts for $20 each (plus freight) and says that 100 per cent of the money is going to New Orleans' Bravest, a fund set up by New York fire fighters to help their brethren in New Orleans. Wiechman notes that he has almost sold out of his first run of 1,000 t-shirts. To order, readers can contact Wiechman at darrylonfire@mts.net.

By: Fire Fighting in Canada. February 2006, pg. 18

CHAPTER EIGHTEEN

WAYS TO STAY MOTIVATED DURING YOUR WORKOUTS

Sometimes the hardest thing about working out is to get the right attitude to make it to the gym. We'd all prefer to be in shape because firefighting requires it. A lot of careers do not require strength, stamina and bravery on such short notice as firefighting. At the sound of an alarm whether it's 4:00 am or in the middle of the day, we are expected to be at the top of our game instantly. Being in great physical shape helps our body keep up with the physical and mental demands of a firefighter.

Choosing a career in firefighting obligates you to be in shape. Many reasons exist why it's important for you to have yourself ready to show you can or have the condition to handle high levels of physical stress. You will be tested on your physical strength, endurance and agility at some point throughout your recruitment process and it has to be there in a crunch. Being in top physical condition will also

Dreams and dedication are a powerful combination.
— William Longwood

aid in preventing injuries to your body on and off the fire grounds. The last thing you need to happen is to get hired by a fire department only to end up hurting yourself, preventing you from making it through your probation period.

Choosing a workout can be challenging and intimidating. There are so many styles and types of workouts available that make it rather difficult to determine what workouts get you the results. Keep in mind that workouts have come a long way in the last 10 years and most of them have the same concepts. They all put physical stress on the body which stimulates adaptation and muscle growth. Exercise will boost metabolism, shed extra pounds, build lean muscle, provide daily-living energy, give you time for yourself, relieve stress and body tension, and re-educate muscles that have become weak from injuries or just plain lack of use. You must have endurance.

For the best results you must have a routine schedule that will best suit you and your level of experience as an exerciser. You have to be mentally as well as physically ready to move to the next level. Advancing too quickly could cause you to become discouraged or cause injury. When working out it's important that you concentrate to enjoy your routine because if you don't eventually you will start making excuses why you can't work out just to avoid it. My advice is if you don't like your workout find another one. Choose carefully from the different types of workouts available to get the right one and the results you want. Consult a trainer.

As a firefighter you must be in great shape, every time you report for duty. The problem is, how do you stay focused to achieve it? Let's look at some ways to stay motivated during your workouts.

Set Goals: Long range and short range goals help keep you on track and define what you what to achieve. Here are some things to keep in mind that will help you achieve your goals.

- Make sure your goals are measurable: A goal like "I want to be strong" is too unclear and hard to measure. But if you say "I want to bench press 275 lbs for 6 reps by my fitness test on April 6th" you are stating a measurable goal.
- Don't set goals that are unattainable: Be realistic. Frustration occurs when unrealistic goals are made and chances are you will want to give up. On the other hand if goals you have set are too easy for you to achieve chances are you won't see the results you want and find a reason to give up.
- To achieve your long term goal. Set short term goals to give yourself the necessary stepping stones: For example, if your long term goal is to squat 300 pounds you should set some short term goals like increasing your weight weekly or monthly by five or ten pounds until you reach 300 pounds. It's a lot easier to increase the weight you're carrying by five pounds per workout to achieve your goal than increasing it by seventy-five pounds on your first time. Take your time to build muscle slowly.
- Make your workouts enjoyable: If you set your attitude to have fun while you work out you are less likely to fall short of your long term goal.
- Add variety: You can lose concentration if you become bored with your workout. If there is a particular exercise you don't like, vary how often you do it or the amount of reps you do. There are many different machines or exercises that will produce the same result. Don't limit yourself to just one way of properly completing each exercise.
- No one said you have to work out alone: Find a friend to work out with. Make sure they take it as seriously as you do or you will have problems keeping them motivated as well as yourself.
- Be careful to recognize you may become bored with your workouts: You may feel that you've reached a plateau. This is a normal part of achieving your goal. You may have to adjust up your exercises, sets, and number of

repetitions. Research new workouts all the time: you may find one that better prepares you for firefighting compared to the one you're currently doing.

- Working out must become a serious routine: By scheduling workouts it becomes a fixed daily activity. Try to work out the same time every day. This will help because you will eventually feel unnatural, if you don't do your workout at that scheduled time. This will help motivate you whether you know it or not.
- Get serious. Keep a record. Make a chart to record your achievements every day. Record time, rep's and weight. You'll soon see your objective being met and your good feeling and sense of accomplishment will be most rewarding.
- Remember planned sleep and diet is part of creating endurance physical growth.

Figure 4–3

Make copies of this workout log and use it to keep track of your workout performance.

DATE:		WEEK:								
EXERCISES	**SET 1**		**SET 2**		**SET 3**		**SET 4**		**SET 5**	
DAY	WEIGHT	REPS	WEIGHT	REPS	WEIGHT	REPS	WEIGHT	REPS	WEIGHT	REPS
1.										
2.										
3.										
4.										
5.										
DAY	WEIGHT	REPS	WEIGHT	REPS	WEIGHT	REPS	WEIGHT	REPS	WEIGHT	REPS
1.										
2.										
3.										
4.										
5.										

! KEY POINTS TO REMEMBER

- Write as many written aptitude test as you can "Perfect practice makes perfect".
- Fire departments sometimes create their own written aptitude test. Most use external test writing agencies.
- Read over the entire test before you begin to write it.
- Get lots of sleep the night before the test.
- Be sure you know exactly where your test is taking place; being late is not an option.

NOTES

_____

_____

_____

_____

_____

_____

_____

_____

_____

_____

_____

_____

_____

NOTES

Part V

Interview Stage

CHAPTER NINETEEN

THE INTERVIEW

Fire departments use interviews to help determine which applicants have the personality and characteristics they're seeking. Interviews are not intended to verify the skill level of the applicant. In most cases your accomplishments such as fire experience, education, work experience, etc., are the reasons why you were given the opportunity for an interview. Interviewers' main purpose is to assess your interpersonal skills, integrity, communication skills, judgment, decision making, abilities, respect for diversity and adaptability. Interviews also give you an opportunity to assess and learn about the fire department and the city where you are applying.

Fire Departments look for:

- Over-all professional attitude and appearance
- Someone who is well put together; meaning physically fit, healthy, self confident.
- Maturity and composure, someone who can handle responsibility with life experiences to back it up.
- Great interpersonal skills, modesty, and an over-all team player attitude.

Never underestimate the power of professional appearance.

- Sincerity and compassion towards being a firefighter.
- Strong work ethic, hands-on skills and mechanical aptitude.
- Understand the importance of following orders, guidelines and procedures.
- Demonstrated leadership in projects and organizations.
- Communication skills.
- Motivated to encourage others to do well; resolve complicated problems in a team environment.
- Able to adapt to new environments and settings.
- Capable to assume high levels of responsibility at a moment's notice.
- Up-standing citizen who has respect for the law and people.

WHAT IS AN ORAL BOARD?

Fire department and city human resources personnel have been chosen to select the ideal candidates from the applicants for the available job openings. Oral boards have a number of representatives who are usually decided by the municipality involved. In larger oral board settings, it is easy to be discouraged and intimidated. Don't! When you know what is expected of you and by undertaking a few simple preparations, you can make a more favorable impression and minimize any nervousness you may feel and show.

CHAPTER TWENTY

HOW TO PREPARE FOR THE INTERVIEW

The best preparation for the interview is to practice using mock interviews with someone who can be helpful. Have someone you know (or ask someone from your city's human resources department) to ask you questions that you can practice answering in an interview setting. Use the interview questions provided in this chapter as a guide. Through repetition your confidence will increase for two reasons:

1. You will be better prepared through studying well.
   (There'll be fewer pauses using "Uh,")

2. You will be more at ease and confident in this type of practice
   environment or setting.

Candidates spend hours increasing their strength and fitness level and spend many hours volunteering and taking courses but, may only spend a short time preparing themselves for the interview. I guarantee, if you go into an interview just 'winging it'—you will not do as well as you could have, had you prepared mentally.

The reason you must consider every answer during the interview is you will be nervous and we all know what nerves can do to us. I'm not suggesting that you memorize your answers word for word. However, be sure to review the example interview questions in this book and also review the night before so you have a generalized answer to each question. For example, you know "why you want to become a firefighter", but do you know "why you would make a good firefighter?" I guarantee if you hesitate telling about why you would make a good firefighter the oral board won't be sure either. By already knowing what you're going to say, and being prepared by rewriting the good reason why, you have just reinforced in your mind that you will confidently give the correct answer. Not only will you feel confident, you have answered the questions properly but, the oral board is bound to be impressed as well. Don't be in a hurry they want you to succeed.

www.becomingafirefighter.com

Become a member and use the "Mock Interview" feature. Available exclusively to members the mock interview is designed to randomly ask you interview questions taken from a question bank of over 50 interview questions commonly used in fire department interviews.

WHAT TO WEAR TO YOUR INTERVIEW?

Your interview is a very important part of the entire recruitment process—perhaps the most important. Clothing for men would be conservative business clothing such as suit, dress shirt, tie and dress pants; don't wear jeans and sneakers. Women should wear conservative business clothing as well, applicants with long hair should put their hair up for a more professional appeal. Whatever you do, don't attend the interview wearing old wrinkled dress clothes dragged from your closest. This is a crucial moment in your life. You have already spent thousands of dollars on schooling and courses, so spend what's needed on clothing so you feel confident about yourself. With proper business attire and a confident professional attitude, the oral board will realize that you really want the job.

HOW TO HAVE A SUCCESSFUL INTERVIEW

You have to imagine what it is like to be one of the interviewer's on the oral board. It has been a long week, interviewing hundreds of candidates, and much to their relief in walks the final applicant—it's you. The oral boards' first impression about you is positive because you are well put together, you have a nice smile and will allow that you look a little nervous. They have already anticipated you to be like everyone else interviewed to this point. It is now your job to make a different impression. Statistics have proven you have thirty seconds to catch the interviewers' attention and if you don't that person will put their mind on autopilot until you are done your interview. A way to avoid this from occurring is by not overdoing your opening statement. Fire departments like to use the question "Tell us about yourself." This is supposed to be an easy ice breaker question to help you relax your nerves and to adjust. Your answer should be no more than one minute in length. Talk about your family or hobbies, keep it simple and brief. You don't want to overdo it. If you spend fourteen minutes of the twenty minutes interview talking about your hobbies, then how is it possible to get through the twenty remaining questions they have to ask you in six minutes?

☐   Get a good sleep the night before.

☐   Stay positive and keep reassuring yourself that you're doing fine.

☐   Keep doubts to yourself.

☐   Nobody knows you better than yourself.

☐   Try to use rejection as a positive thing, learn from your mistakes.

☐   Keep your confidence up.

☐   You have the upper hand; the interviewer knows less about you than you do.

☐   It's expected that you'll be nervous, don't sweat it. Concentrate on getting the right answers not how you feel standing in front of a selection board.

☐   Avoid sweaty palms by sitting with your palms exposed to the air. Put them on your knees. Raise them for emphasis.

☐   Combat nervousness by deep breathing before the interview sparingly by relaxing your mind and muscles.

☐   Mind and body act together. Relax your body and your mind will be more tuned in to the interview.

☐   Do only a light review when preparing the night before it's all you need if you have spent your time preparing properly in the prior weeks.

CHAPTER TWENTY-ONE

HOW TO SURVIVE THE INTERVIEW

Do

- Prepare a list of your qualifications, experiences and personality traits. Refer to this list when selling yourself in the interview.
- Concentrate on general questions the interviewers might ask such as "Why do you want to be a fire fighter?", "What can you offer?" and "What are your strengths and weaknesses?"
- In case you have to clarify a section of your résumé make sure you know it inside and out so you can answer a question without hesitating.
- Do thorough research. Write everything down and organize it to be easy to find. Make sure you have thorough background knowledge about each fire department interviewing you including their fundraising for charities and all events in which they are involved, etc.
- Know your strengths. Use the minimum qualifications list of the recruiting fire department as a reference. Write down experiences you have that match each job description requested.
- Know your weaknesses and tell how you are working to improve them. Fire departments like candidates who can acknowledge their weaknesses. Be prepared to provide detailed answers; try to give examples that are

positive such as "My friends would say that I spend too many weekends taking courses and not enough time having fun" This is a lot better answer than saying "I have a difficult time following instructions when I am excited."—Notice the difference in the two responses. The first example is harmless; the second example is a concern to the fire department—its damaging to your career, be sure not to make this mistake.

Don't

- Don't bring a copy of your résumé to the interview. If you want everyone on the oral board to have a copy of your resume, arrive well in advance and give copies of your résumé to a secretary for the interview panel to receive them. Make sure *www.becomingafirefighter.com* reviews your résumé and cover letter.

DURING THE INTERVIEW

Do

- Give a firm handshake and smile when introductions are shared. Look each person in the eye and don't look away. Be genuinely interested in meeting the interviewer. Think of him/her as an old pal, but don't show it.
- Folding your hands in your lap will help you maintain proper posture during the interview. Looking attentive and sitting-up straight makes a lasting impression no matter where you are it demonstrates pride in your-self and your career.
- Be polite and courteous.
- When speaking change the pitch of your voice, don't use boring mono-tone.
- Listen to the whole question and let it sink-in before you attempt to answer.

- If you need to hear the question again or if you didn't fully understand the question the first time, ask the interviewer to please repeat the question. Not only will this give you more time to answer the question, but now you have heard the question twice and in two different ways. Be careful—you have only twenty minutes.
- Ask a question or make a statement to clarify a question. The question may be: "You see your partner putting a tool in his pocket, what do you do?" Ask yourself, "What is the root of the question?" Are they telling you that your partner is stealing? Don't assume anything—ask a question, "Is the tool his?" They might say, "No, he's stealing it," even though you have now scored the points. I would then say, "Stealing is wrong," and if he won't put it back, you might say, "Let's go together and talk to our captain, because removing tools or equipment without approval is stealing and stealing is against the law and the rules of the department."
- Watch your mannerisms and language. Don't chew gum, tap your foot, jiggle your foot, or show impatience.
- Be friendly to office staff, such as the receptionist or anyone you meet before or after the interview.
- Make direct eye contact with everyone on the oral board. However, looking away, or at your notes to ponder a question for a moment is acceptable.

Don't

- Don't use first names (unless requested) "Sir or Ma'am" demonstrates that you are polite.
- Don't smoke (even if invited).
- Don't sit down (until invited).
- Don't show anxiety, boredom or monitor your watch.
- Don't always read from your notes.
- Don't use a pen or any writing implement that could distract yourself and the interviewers.
- Don't provide unnecessary (too much) information.

- Don't ask to discuss salary, vacation or benefits—bad timing! These topics you should already know by researching properly.
- Do not ramble. Speak loudly and clearly. Make the effort to answer each question wisely, but be brief.

DEPARTURE OF THE INTERVIEW

Do

- When the interview is coming to an end, say 'Thank you' for their time and questions regardless of how you feel the interview went. Always appear confident. Don't show disappointment.
- Send a "Thank you" letter addressed to each individual of the interview board within the next day following the interview. Showing consideration and keeping in touch can make the difference of being recruited.
- Ensure when you're interviewed by more than one person, you have the correct spelling of the names of every individual who interviewed you. It demonstrates your consideration and is critical for the follow-up thank you letters.
- Always depart in the same polite and assured manner that you used to enter the interview.

Don't

- Don't reveal that you are discouraged even if you felt the interview did not go well; keep it to yourself even when you've exited the building. You never know who is watching your reactions as part of the process never ever express anger anywhere.
- Don't be overly confident if you felt the interview went well; be grateful. Keep it to yourself, it shows maturity.

CHAPTER TWENTY-TWO

INTERVIEW PREPARATION DETAILS TO KNOW

It is vital to take the time before your interview to research general facts about the city and fire department. Having this knowledge greatly improves your chances of having a successful interview.

City

- Mayor's name
- Population of the city (winter, summer, spring, fall)
- What are the city attractions? Uniqueness?
- Natural hazards? (Escarpments, water, etc.)
- Breakdown of city (industrial, commercial, residential)
- Expansion or annexing in future?
- Number of hospitals and locations
- Airport, train and bus stations, subway, etc.
- Man-made hazards (Chemical plants, mines, etc.)

Most cities have a Tourism Information service. Take the time to contact this resource and have them send you any information regarding the city including maps, brochures, contact numbers, info on real estate, shopping, schools, services, festivals, and so on.

Fire Department

- Chain of command
- Chief's name
- Deputy chief's name
- Number of fire halls and their location in the city (try to visit each one so you can commit locations to memory and neighborhood served)
- Type of shifts worked
- Charities in which they're involved
- Specialized teams (high angle, hazardous materials, water rescue, etc.)
- Number of calls annually (at each hall, if possible type)
- Daily station duties
- Salary, benefits
- Public education involvement (home safety visits)
- Mutual or automatic aid policies
- Division of jurisdiction regarding fire districts
- Number of firefighters employed by the city
- Department history (F.D website, city hall, library)
- Composite fire department or integrated EMS (Firemedics)
- Future growth (new fire halls, new trucks, etc.)
- Have they had any major fires or incidents lately (try to visit sites)

SAMPLE INTERVIEW QUESTIONS
General or Traditional. Common questions about yourself.

If you are serious about becoming a firefighter write a full answer to each question.
- What are your positive personality traits?
- Name all of your strengths
- Why do you want to be a firefighter? (set in priority)
- Why do you want to be a firefighter for the city of _____?
- Why should they hire you?
- What computer skills do you possess?
- Do you have a criminal record?
- What is the definition of stress to you?

- How do you deal with everyday and unanticipated stress?
- What is one word your friends would use to describe you?
- What have you done to prepare for this interview?
- What have you done to prepare for a career in the fire service?
- Where do you see yourself in 5 years? 10 years?
- What is the most appealing aspect of being a firefighter?
- What is the least appealing aspect of being a firefighter?
- What do you consider to be your strongest asset? Your weakest?
- What activities have you been engaged in, in order to improve/maintain your levels of fitness?
- Why would you be a good firefighter?
- Why have you chosen to be a firefighter instead of a police officer, ambulance paramedic, etc.?
- What do you believe to be the most important attributes of someone being a firefighter?
- What are your plans if you are not successful with your application?
- What physical demands would you expect to be placed on you during firefighting operations?
- Do you have any experience working in a team environment and have you held any responsibility within the team/group?
- What do you believe are the characteristics of a good balanced team?
- Define sexual harassment and give your feelings about the subject?
- What would you do if you were a witness to someone being sexually harassed?
- Honesty and integrity: Define them and why are they important in the fire service? (Use a dictionary to be accurate)
- The fire service has an equal opportunities policy. What do you understand by the term Equal Opportunity?
- The role of a firefighter involves working unsociable hours, working night shifts, bank holidays etc. How will you cope with this disruption to your family social life?
- What other special services does the Fire Department offer the public?

TIP!

If the opportunity arises to ask any questions about the interview it is a good idea not to say very much or anything at all, but make a closing statement to solidify your enthusiasm and abilities for the job.

- How do you think you would cope when faced with casualties and/or fatalities at operational incidents?
- Pride and loyalty: define them and why are they important in the fire service?
- You are ____ years old. Why have you left it so long to apply for a career in the fire service?
- If you were asked to choose a hero, who would it be and why?
- Describe teamwork.
- What do you think you will dislike about this career?
- What does diversity mean to you?
- What three words describe your personality?
- How long have you wanted to be a firefighter?
- What made you decide to be a firefighter?
- Do you have any problems with relocating?
- Do you have any family members who are firefighters?
- How many interviews with other departments have you had?
- Have you ever done a physical testing before? How did you do?

Work Habits

- What would your previous employer say about your work habits?
- What duties do you perform at your present job?
- Have you ever abused sick days at work?
- If your shift started at 8:30 am what time do you think would be considered late?
- What type of person would you find most difficult to work with?
- Would you ever disobey an order? If so, when and why would you disobey an order?
- How would you feel taking orders from somebody younger than yourself; can you demonstrate a time when you've been in this situation?
- How will you feel taking orders from a female, can you demonstrate a time when this has happened?
- If we were to contact your present or former employer, what would be one negative point they might raise about you?
- Attending operational incidents is an exciting and often rewarding

experience; however, a significant part of a firefighters is spent on routine duties such as cleaning and maintaining equipment, etc. How do you think you would cope with the routine aspects of the job?

- Give some examples of any jobs you have done of a practical nature either at home, at school, or at work.
- Firefighters are required to use their hands in their work, what examples can you provide of any practical tasks you have been involved in?
- Our fire department can not tolerate casual absences or convenient absences due to sickness. Why do you believe the fire department is so strict with its absentee policy and what are your thoughts on this matter?
- What single trait must every firefighter possess?
- What will you bring to this department?
- What do you think the mission statement should be for our department?

Qualifications

- Relate your education and work experience to firefighting.
- Do you think you are qualified enough to be a firefighter?
- What skills do you have that would be a benefit to the fire department?
- What fire related training do you have?
- What responsibilities do you have in either your job or domestic activities?
- Have you any experience of communicating with the general public? Give some examples. What do you consider the most important factors in being able to communicate clearly with the public or group of people?
- One aspect of a firefighter's role is to produce written reports, would this present any problems or concerns?
- How do you arrive at making difficult decisions?
- What practical medical experience do you have?
- What was your reasoning for attending _____ College?
- Have you ever handled an emergency situation?
- Have you ever experienced a dangerous situation, if so, how did you cope? What were your thoughts and what were your priorities?
- What did you like about being a volunteer firefighter?
- How many calls a year do you run?

Fire Department Knowledge

- What do you like about our city?
- What do you like about our fire department?
- How many calls do we run annually?
- What medical training do you require to work here?
- What roles do our firefighters play in the community?
- What is our chief's name?
- Have you ever visited any of our fire stations?
- Do you know how many fire stations we have?
- When would you disobey an order?
- What special teams do we have?
- If hired what benefits would you add to our department?
- What is the typical daily routine in a firehouse?
- What will you do with your spare time while on duty?
- What duties does a firefighter perform?
- What is the most essential duty a firefighter performs?
- What do you think the future holds for the fire service?
- In what direction do you see the fire service going?
- How could you help maintain good relations around the firehouse?

Behavioral—These questions are designed to give fire departments an insight regarding each applicant's frame of mind such as negotiation, persuasiveness, teamwork, communication skills, decision making, problem solving, planning, organization, and coping with pressure.

- Tell us about a time when you had to deal with an angry or hostile person (e.g. a customer, boss, co-worker, and friend).
- Tell us about a high stress situation you have experienced.
- Tell us about an instance where you helped someone you didn't know.
- Tell us about an instance where you were responsible or accountable for someone else.
- How would you react to unwanted attentions from your work mate (i.e., bullying, name calling, initiation ceremonies, etc.)?

- Can you tell us about a time when you have told a lie and had to or chose to go back and confess?
- Tell us about a time when you had to exercise self-discipline.
- Can you explain the difference between discipline and self-discipline?
- Discipline is considered to be very important to the fire service, why do you think fire departments promote discipline?
- Can you tell us about a time when you've been caught in an embarrassing situation? How did you overcome this?
- Tell us about a time when your positive attitude influenced others to be positive or motivated.
- Confidentiality is critical in the fire service, why do you think it is so critical?
- How would you handle a conflict with a co-worker?
- Tell us what you would do if you caught a co-worker stealing?
- Tell us what you would do if you witnessed a female co-worker being sexually harassed by male co-worker.
- Tell us about a time when you influenced someone in a positive way.
- Tell us about a time when you had to stand up for someone.
- How would you handle working with a negative co-worker?
- Give us an example when you took the initiative at work.
- Tell us in your own words the significance about rank and role.
- Give us a time you disagreed with your supervisor and how you handled it.
- Tell us about any instance when you worked with others in order to accomplish a common goal.
- Tell us about a time you experienced taking reasonability for your actions.
- Give us an example of a time when you were involved in an emergency situation. Describe the situation and how you handled it.

Case or Hypothetical—Defines your problem-solving skills at a moment's notice by giving examples of what you'd do "if".

- We hired you and your home town soon announces they are hiring. What would you do?
- You are at a fire call and your captain asks you to climb a ladder you know is unsafe. What would you do?
- A co-worker member is making racist comments and it's bothering you or a co-worker member. What would you do or say?
- Your captain takes you aside and you notice he has been drinking. What would you do?
- You are walking past a house and smell gas and you notice what looks like someone unconscious in the house. What would you do?
- You are evacuating an entire floor of an apartment building because of a small fire on the floor below. You encounter a hostile occupant who refuses to leave his apartment. How would you handle this situation?
- You have just been informed about a co-worker who has been accused of committing an offence. What would you do?
- While you are exiting a building and you encounter an individual who appears to be unconscious on the sidewalk. What is the most important thing you would do?
- You are at an off-duty party and notice your captain smoking drugs. What would you say or do?
- You have just returned to the fire hall from a life-threatening incident when you notice a relative of the victim waiting there. How would you approach this situation and what would your actions be?
- You arrive at a structure fire, which appears to be puffing smoke from its windows and doors and has been burning for a while. Your Captain instructs you to ventilate the roof. What would you do?
- You arrive at a structure fire and notice that the building looks abandoned and the roof is showing indications that it is ready to collapse. Your Captain orders you to advance a hoseline to the front door and prepare to enter the structure. What would you do?

Practical—Fire

- What is a flashover?
- What is a backdraft?
- What are the signs of building about to collapse?
- What are the types of building collapses?
- What are the classifications of fire, the associated letter and explain how each extinguishes that fire?
- Describe the different methods of heat transfer.
- Explain the fire tetrahedron.
- Name the classifications of building constructions.
- What are the steps you would take at the door before entering a structure on fire?
- Describe the difference between a wye and a siamese.
- What are the different categories of skin burns? Explain.
- Explain the chain of command.
- What are the different stages of fire?

Practical—Medical

- What does ABC stand for in CPR?
- What is the rate of respirations for an adult, child, and infant?
- What is the average resting heart rate for adult, child, and infant?
- Name three different types of fractures.
- What are the indications of a head injury?
- What does OPA stand for when it comes to human airways?
- What is the number of chest compressions for an adult during single rescuer CPR?
- How many breaths per second during CPR for a child?
- How do you treat a person suffering from heat stroke?
- What does BVM stand for?

FIREFIGHTER GOES THE DISTANCE DRIVING RIG ACROSS CANADA FOR DONATED GEAR.

A Toronto firefighter is driving a rig across Canada today in hopes of filling it with firefighting equipment for poor countries. So far Ron Kyle, of Barrie, has collected about $200,000 worth of firefighting equipment from Toronto Fire Services, Central York Services, and now he's on his way to the Calgary Fire Department to pick up another load.

By: Tracy McLaughlin. *Toronto Sun*. Friday September 3, 2004

CHAPTER TWENTY-THREE

COVER LETTER, RÉSUMÉ, REFERENCE LETTER AND FOLLOW-UP LETTER

Cover Letters: A cover letter should always accompany your résumé and is used whenever e-mailing, faxing, or mailing your résumé to fire departments. You can submit your résumé and cover letter to any fire department whether or not they are actively recruiting.

A cover letter is designed to give a brief indication of the most relevant skills you have without plagiarizing the job ad. Cover letters give you an opportunity to demonstrate your writing skills as well as give the fire department an overview of your achievements. Studies have proven that well-written cover letters enhance the chance of a résumé being read by up to 65%. Try to refrain from stating the obvious such as; you are interested in being a firefighter. The cover letter must be able to stand out on its own even though it always accompanies a résumé. Do not send it hand written. Use a computer printout and keep a copy.

COVER LETTER WRITING TIPS

- Format your cover letter to resemble a typical letter you would send any employer. All text such as your name and address and the recipient's name and address should be aligned on the top left margin of the page. Example,

John Doe

224 Ernest Ave.

London, ON

W3E 4W3

(555) 000-0000

_____ (Recipient's Name)

Human Resources Department

City of Toronto

City Hall, 1st Floor

500 Front Street

Toronto, ON

N6V 3T3

- The actual letter should be in short paragraphs and never more than one page in length, including your signature.
- Always find a specific individual you can address your cover letter to rather than a general recipient such as the "Ottawa Fire Department" or to Human Resources Representative.
- Cover letter should always be dated and signed in ink.
- Use lower and upper case letters. NEVER USE ALL CAPITALS.
- It is not necessary to list every qualification you have. Remember it is only an overview of your qualifications; always list the most important qualifications you have first.

- Each and every fire department you apply to should receive a personalized cover letter. If a fire department lists in their recruitment ad they are looking for recruits that possess a certain quality or qualification— and you have it, be wise to say so in your covering letter first.
- Don't send a cover letter that contains any typos, misspellings, incorrect grammar or punctuation, smudges or white outs. It must be perfect. Have someone else read before sending to double check.
- Use simple language, uncomplicated sentence structure and short sentences.
- Avoid any negativity.
- Always personally sign the letter. Type your name beneath your signature.

Example,   *John Doe*
              John Doe

- Use quality paper.
- Only use black ink, no colour.
- No pictures or diagrams.
- Don't overload with the constant use of "I".
- Avoid slang, profanity and clichés.
- Use short paragraphs (4 or 5 lines).

SAMPLE COVER LETTER

John Doe
224 Ernest Ave.
London, ON
W3E 4W3
(555)000-0000

September 04, 2006

_____ (Recipient's Name)
Human Resources Department
City of Toronto
City Hall, 1st Floor
500 Front Street
Toronto, ON
N6V 3T3

Dear_____,

    Thank you for the opportunity that allows me to apply for the Probationary Firefighter position. I am confident in my ability to achieve the standards required by the Toronto Fire Department.

    I have recently completed NFPA 1001 fire fighter level one and two programs and the NFPA 472 hazardous materials operations level at _____ Fire School in _____ (Province/State). I feel that my education and experience coupled with my dedication to work in the fire services make me an ideal candidate.

    In addition to graduating from _____ Fire School, I have successfully graduated from _____ College's paramedic program. Since graduating I have been a certified paramedic for _____ years. I am an outgoing, optimistic, individual with a positive attitude that would be a great addition to any one of your teams.

    An opportunity to discuss your requirements and my qualifications in a personal interview would be greatly appreciated. I am available at your convenience.

Sincerely,

John Doe

John Doe

Résumés: Your résumé is a tool with one specific purpose: to convince the fire department to give you an interview. A résumé is an advertisement; nothing more, nothing less. A great résumé doesn't just tell the fire department what you've accomplished but makes the same assertion that all good advertisements do: If they buy this product (you), they will get these specific, direct benefits. It presents you in the best light and demonstrates to the fire department that you have what it takes to be a successful probationary firefighter.

It is vital your résumé is pleasing to the eye and easy to follow so that the reader is enticed to pick it up, read it thoroughly, and think of you in a positive way. Its purpose is to entice the reader to want to meet and learn more about you. Take the time to create a masterpiece because it gives great insight about you to the recruiters prior to meeting you in person.

RÉSUMÉ WRITING TIPS

- Be sure to limit the amount of pages used for your résumé. Two pages are ideal but, if three or four pages are absolutely necessary then do so.
- Always put your name, address, postal code, phone number, and email address and page number at the top of each page. This way if the pages get separated the author is easily identified.
- Always list courses, education, volunteer work, etc. in reverse chronological order. (Most current to least current)
- Be sure to bold your headings, so they will stand out.
- Do not use italics on your résumé.
- Use upper and lower case letters—NOT ALL CAPITALS.
- Do not list months or days when listing course or jobs only the year(s).
- Never switch tenses, refer to your current position in the present tense. All other entries should be written in the past tense.
- Use a proper type, and open style size (nothing below a 12 point).
- Use at least 1.5 inch margins on each side.
- Use point-form when describing job duties or education, background, etc. Limit lengthy descriptions (no more than five).

- Properly order your résumé headings as follows:
  PROFILE
  FIRE EXPERIENCE
  EDUCATION
  EMPLOYMENT
  RELATED SKILLS (Subtitle: Licences, Certifications)
  VOLUNTEER EXPERIENCE
  REFERENCES

Note: Every piece of information on your résumé should fall under one of the above categories.

- Don't put more information on your résumés than required. This lessens clustering and will draw more attention to the points you wish the reader to see.
- Be organized, relevant, accurate, concise, and appropriately detailed.
- Include the year for each entry on your résumé. (Use 'Present' if currently in position or course rather than the current date; 2002–Present). This will eliminate confusion in cases where your résumé is not seen for weeks and it becomes unclear if you are still employed in that position.
- Be sure to have continuity between the skills mentioned in cover letter, summary of qualifications and the resume itself. If it's in your cover letter, you need to include it in the résumé, too.
- Do not use colour in any way.
- Include the 'Job Title' in bold and list it first. Keep in mind the job title is more relevant than the company name, although the company name must be there.
- Be relevant! Use the minimum qualifications list to determine exactly which skills are important to the fire department hiring and use point form to emphasize those skills in each of the work or volunteer experiences you have.
- Include all types of relevant experience: employment, fire training, co-op placements and volunteer work.
- Take the time before you put a pen to paper and view other firefighting résumés and compare them with each other. Pick out things you like

about them—it could be the format, layout or the text. When you are confident you know what you are doing start your résumé. Use wide margins don't crowd the bottom, it looks better.

- For each job, your responsibilities, accomplishments, and skills, demonstrate how or why they are relevant and transferable; or define the results of what you have achieved.

- You should also colour photocopy all of your certificates and courses that you have taken to submit with your résumé package. It is acceptable to use a scanner to complete this. It is a minor thing but you would be surprised how many applicants just photocopy their certificates in black and white.

- Do not make assumptions about the reviewers' knowledge. Nothing is obvious. The reviewer should not be trying to decipher which skills you may or may not possess. Be very clear. Spell everything out but keep it easy to understand.

- Never include your picture on your résumé page, or submit a photo. Age, ethnicity, appearance are not qualities you want to be judged.

- Never include your hobbies, interests, family information. Nice as these are, they do not belong on a résumé! However if you have valuable information such as you once played professional hockey, I would list this information under 'Profile' category.

- If you attended a fire college, it's a wise decision to photocopy and submit all your transcripts with your résumé package and any awards. It doesn't hurt to show-off impressive grades you've earned. The transcripts should be located in your résumé package between your firefighting diploma and general certificates.

SAMPLE RÉSUMÉ

224 Ernest Ave. / London, Ontario / W3E 4W3 / (555) 000-0000

PROFILE

Summary of Qualifications
- Played professional hockey—Calgary Flames
- Excellent interpersonal skills with team building qualities
- Able to resolve conflicts while remaining sensitive to others' feelings
- Self motivated learner, constantly striving for improvements, constant reader
- Respond positively to problems while working on creative solutions
- Remain calm under pressure to overcome conflicts and problems
- Mature lifestyle with focus on health and safe work practices
- Compatible with work in adverse environments

FIRE TRAINING

NFPA 1001—*Fire Fighter Level I and II*   **2003**
Fire Etc   Vermilion, AB

NFPA 472—*Hazardous Materials Operations Level*   **2003**
Fire Etc   Vermilion, AB

NFPA 472—*Hazardous Materials Awareness Level*   **2003**
Fire Etc   Vermilion, AB

EDUCATION

ALARM TECHNICIAN   **2004-2005**
Alarm College   London, ON

SPRINKLER TECHNICIAN   **2003-2004**
Sprinkler college   London, ON

SECONDARY SCHOOL DIPLOMA   **1995-1999**
London District Collegiate Institute   London, ON

SAMPLE RÉSUMÉ

224 Ernest Ave. / London, Ontario / W3E 4W3 / (555) 000-0000

EMPLOYMENT

PRESENT PATIENT TRANSFER TECHNICIAN   **2004-Present**
Patient Transfer   London, ON

ELECTRICIAN APPRENTICE   **2003-2004**
Wayne's Electrical Services   London, ON

QUALITY ASSURANCE INSPECTOR   **1994-2003**
Boltons Manufacturing   St. Thomas, ON

LABORER   **1994**
City of London   London, ON

RELEVANT SKILLS

LICENCES:
- **B-Z License**—Fanshawe College   **2006**

CERTIFICATES:
- **W.H.M.I.S**—Patient Transfer   **2006**
- **Basic Trauma Life Support—Pediatric**—Toronto EMS   **2006**
- **Basic Trauma Life Support**—Toronto EMS   **2005**
- **CPR Level C**—Life Support Services   **2005**
- **Semi-Automated External Defibrillation**—Life Support Services   **2004**
- **Medical First Responders**—Canadian Red Cross   **2004**
- **Standard First Aid**—Canadian Red Cross   **2004**

VOLUNTEER EXPERIENCE

- **Emergency Responder**, Red Cross, Hamilton Branch   **Present**
- **Firefighter Combat Challenge**, Hamilton, ON   **2006**
- **St. John Ambulance**, London, ON   **2005**

References available upon request

SAMPLE RÉSUMÉ

142 Lundy Lane • London, Ontario • N5R 8L7 • (521) 257-7564 • Jdoe@whatever.com

SUMMARY OF QUALIFICATIONS

- Dependable team player
- Willing to assume all levels of responsibility
- Excellent communication and interpersonal skills
- Thorough understanding of electrical and mechanical components
- Ability to organize and prioritize projects in a fast-paced environment
- Excellent manual skills and technical aptitude
- Conditioned to working in extreme cold and hot ambient environments

FIRE EXPERIENCE

VOLUNTEER FIREFIGHTER   **2004–Present**
Penticton Fire Department   Penticton, ON

FIRE SCIENCE AND TECHNOLOGY   **2004**
Lambton College   Sarnia, ON
- Co-op, Toronto Fire Department (420 Hours)   2005
- Co-op, Sarnia Fire Department (460 Hours)

PRE-SERVICE FIREFIGHTER CERTIFICATE   **2005–2006**
Lambton College   Sarnia, ON
- Co-op, St. Thomas Fire Department (360 Hours)   2004

EDUCATION

REFRIGERATION AND AIR CONDITIONING PROGRAM   **2002–2003**
Mohawk College   Hamilton, ON

ONTARIO SECONDARY SCHOOL DIPLOMA   **1997–2001**
_____ Secondary School   London, ON

SAMPLE RÉSUMÉ

142 Lundy Lane • London, Ontario • N5R 8L7 • (521) 257-7564 • Jdoe@whatever.com

EMPLOYMENT

HEATING, VENTILATION, AIR CONDITIONING   **1994-Present**
AND REFRIGERATION MECHANIC

Doe Air Conditioning and Heating Company   London, Ontario
- Completed five year apprenticeship for air conditioning and refrigeration
- Health and safety representative
- Respond to natural gas and propane leaks as well as carbon monoxide alarms
- Installation, repair and maintenance of positive displacement pumps, centrifugal pumps, motors, air conditioning, heating and refrigeration equipment
- Welding and soldering (arc, brazing, oxy-acetylene, acetylene)

RELEVANT SKILLS

LICENCES:
- Air Conditioning and Refrigeration Mechanic (Inter-provincial)   **2004**
- Gas Technician II   **1999-2004**
- Class B Licence (includes C, D, E, F and G)   **2004**

CERTIFICATIONS:
- Automated External Defibrillation   **2006**
- BTLS Level 1   **2006**
- CPR and First Aid Instructor   **2006**
- Air Brake Endorsement   **2006**

VOLUNTEER EXPERIENCE

- St. John Ambulance Brigade (3 to 4 hours/week)   **2005-Present**
- Big Brothers of London   **2005**
- Assistant Coach for Tykes Baseball Team   **2002**

References available upon request

References: When you are in the process of selecting your references, give it a lot of thought. Selecting too fast at this stage in the game could be detrimental to everything you have worked so hard to achieve. It is critical you control the things you can such as references to ensure your success. The best advice I have for you is to choose those who offer to be a positive character reference for you, opposed to those you have to ask who might be indifferent or opposed to what you are doing. The reason for this is because the individual who has offered is obviously keen on speaking about you and will probably have more enthusiasm in their voice as opposed to someone you have asked.

If you are concerned that you have weak references take the time to create good ones. Get out and meet new people, visit fire halls and talk with chiefs, platoon chiefs, captains, lieutenants, firefighters etc. It is not hard to meet new people that have common interests as you do. A lot of individuals in the fire service never hesitate to take someone under their wing and help out. Most firefighters remember how hard it was to get a job on a fire department and like and enjoy being your mentor and/or friend helping you out.

Note: If someone does take the time or the interest to help you out make an extra effort to thank the individual promptly and keep in touch.

SAMPLE REFERENCE LETTER

JOHN DOE   Pg. 1 of 1   (Sample # 1)

224 Ernest Ave. / London, Ontario / W3E 4W3 / (555) 000-0000

REFERENCES

Name of reference
Phone #
Title (Employer)
Address
Postal Code

Name of reference
Phone #
Title (Firefighter)
Address
Postal Code

Name of reference
Phone #
Title (Co-worker)
Address
Postal Code

Follow-Up Letters: You may find this surprising but the follow-up letter is extremely important and should not be taken lightly (it isn't by the one who receives it). Consider this scenario: The fire chief and the other oral board representatives are preparing to make their decision. You're one of the applicants in the running for a firefighting position and the chief is having a tough time making a decision. That morning your professionally prepared follow-up letter arrives and reconfirms your interest in the position. Guess who most likely gets to the top of the job offer list?

You should always send a follow-up letter when you receive a rejection letter or if either the chief or human resources representative responds to your employment inquiry. This keeps communication open and your name active.

FIREFIGHTERS HONORED FOR BRAVERY

The Ontario government recognized two firefighters and three police officers for bravery during a ceremony held Feb. 2 at Queen's Park.

James K. Bartleman, Lieutenant Governor of Ontario, was joined by Community Safety and Correctional Services Minster Monte Kwinter to present the medals. These medals are the province's highest honor in recognition of firefighters and police officers whose actions demonstrate outstanding courage and bravery in the line of duty.

"The heroic actions of firefighters and police officers remind us of the risks they face every day." said Bartleman. "We owe them our gratitude and our thanks for all they do to keep us safe."

The recipients of the 2005 Ontario Medal for firefighter bravery are Capt. Ralph Novel of the Mississauga Fire and Emergency Services, and Lieut. Mark Smith of the Sarnia Fire Rescue Services.

On December 21, 2004, Capt. Ralph Noble was walking his dog in a park area near Lake Ontario when he heard a faint cry for help. He saw a woman in the water about 35 meters from the shore. Noble caught the attention of a bystander and asked her to call 9-1-1. He then took off his jacket, unleashed his dog, and entered the icy waters to rescue the woman by throwing the leash towards her. Despite the victim's advanced stage of hypothermia, she held on to the leash and Noble pulled her in, saving her life.

An independent body of citizens representing all regions of Ontario determine medal recipients. A total of 163 Ontario Medals for Firefighter Bravery have been awarded since 1776.

By: Canadian Firefighter and EMS Quarterly. April 2006, pg. 8.

STARTING YOUR RÉSUMÉ

Prior to starting your résumé, use this outline guide to assist you to brainstorm and collect the information needed. It's important that you include all vital information. Be sure to list experience, locations, what and when.

FIRE TRAINING: List of Accomplishments

Course/Experience: _____

Location: _____ Date: _____

Course/Experience: _____

Location: _____ Date: _____

Course/Experience: _____ _____

Location: _____ Date: _____

EDUCATION: List of Accomplishments

Course: _____

Location: _____ Date: _____

Course: _____

Location: _____ Date: _____

Course: _____

Location: _____ Date: _____

Course: _____

Location: _____ Date: _____

EMPLOYMENT: List of Accomplishments

Position: _____

Company: _____

Location: _____ Date: _____

Position: _____

Company: _____

Location: _____ Date: _____

Position: _____

Company: _____

Location: _____ Date: _____

Position: _____

Company: _____

Location: _____ Date: _____

RELEVANT SKILLS: List of Accomplishments

Course/Experience: _____

_____

Location: _____ Date: _____

Course/Experience: _____

_____

Location: _____ Date: _____

Course/Experience: _____

_____

Location: _____ Date: _____

```
┌─────────────────────────────────────────────────────────────────────┐
│                                                                       │
│  VOLUNTEER EXPERIENCE: List of Accomplishments                        │
│                                                                       │
│  Organization/Involvement: _____   │
│                                                                       │
│  _____  │
│                                                                       │
│  Location: _____ Date: _____    │
│                                                                       │
│                                                                       │
│  Organization/Involvement: _____   │
│                                                                       │
│  _____  │
│                                                                       │
│  Location: _____ Date: _____    │
│                                                                       │
│                                                                       │
│  Organization/Involvement: _____   │
│                                                                       │
│  _____  │
│                                                                       │
│  Location: _____ Date: _____    │
│                                                                       │
│                                                                       │
│  Organization/Involvement: _____   │
│                                                                       │
│  _____  │
│                                                                       │
│  Location: _____ Date: _____    │
│                                                                       │
└─────────────────────────────────────────────────────────────────────┘
```

Add descriptive words like the ones listed below when explaining the items on your résumé:

- Highly organized and dedicated, with a positive attitude.
- Able to handle multiple assignments under pressure.
- Excellent written, oral and interpersonal skills.
- Thrive on working in a challenging environment.
- Widely recognized as an excellent care provider and patient advocate.
- Outstanding interpersonal and communication skills, superior accuracy in patient history, charting and other documentation.
- Dependable team player.

- Willing to assume all levels of responsibility.
- Excellent communication and interpersonal skills.
- Thorough understanding of electrical and mechanical systems and components.
- Able to organize and prioritize projects in a fast-paced environment.
- Excellent manual skills and technical aptitude.
- Conditioned to working in extreme cold and hot ambient environments.
- Ability to communicate orally, or in writing, or via computer/electronic means.
- Working well with others in order to achieve a common objective.
- Being able to motivate and encourage others.
- Ability to see opportunities and to set and achieve goals.
- Thinking things through logically to determine key issues.
- Analytical and able to handle change and adapt to new situations.
- Ability to relate well to others and to establish good working relationships.
- Competent understanding of numerical data, statistics, graphs, also computers.
- Excellent interpersonal skills with team building qualities.
- Able to resolve conflicts while remaining sensitive to others' feelings.
- Self motivated learner, constantly striving for self improvement.
- Adaptable to change and diversity.
- Respond positively to problems while working on creative solutions.
- Remain calm under pressure to overcome conflicts and problems.
- Mature lifestyle with focus on health and safe work practices.
- Compatible with work in adverse environments.

CHAPTER TWENTY-FOUR

MAKING THE SHORT LIST

A hiring pool is a short list that consists of qualified applicants found and accepted during a recruitment process. These applicants are usually rated or scored to determine who will be hired before others. Fire departments rarely disclose your current position on the short list. Applicants on this short list must continue to take courses and better themselves in order to maintain the privilege of remaining on the short list. Failure to improve your résumé during this stage could be costly and an advantage to other applicants. Once you have made the short list, my advice is to take a week or two, maybe even a month off, to get your thoughts simmer and plan your next step. Then start taking and compiling good courses so you can remind the fire department why you made the short list. Continue to better yourself and move up the list until the day you are recruited. This will prove you're the type of person they have been waiting to recruit.

You will be notified once you're in the hiring pool and informed when an entry-level firefighting position becomes available. Keep in mind just because you're ranked fourth on the short list doesn't necessarily mean you will be the fourth person hired. You may very well be the fourth person the fire department calls. But how do you know the other applicants on the list have not already been hired by other fire departments, or may refuse the job offer due to the fact they're going through the recruitment process for the fire department in their home town?

It is vital that you continue to update your personal file at the fire department. This gives you a legitimate reason to visit the fire hall again and perhaps say "Hello" to the chief or deputy chief and others. The short list has the potential to stay active for one to two years depending on how many applicants remain on the list. Short lists have the potential to deteriorate before fire departments have had the opportunity to hire all the applicants on the list. In this case fire departments will host another recruitment process and create a new hiring pool to fill vacant positions.

When up-dating your personal file at the fire department you should make sure it is done professionally. It is recommended the update letter follow the same profile as your résumé. It's important that your complete name and mailing address is on the update letter as well as specific information about the updated material, show the current date. For example if you have taken another course you should list: the course name, its location and the date of completion. This should be done for any information to be properly added to your résumé. If you have intentions of updating your personnel file on the day of your interview; be sure to give your information to the office administrator prior to the interview and insist they deliver the material to the interview committee. This will be more professional and give the committee time to review the information before the interview starts.

**Courage is what it takes to stand up and speak;
Courage is also what it takes to sit down and listen.**
— Carl Hermann Voss

SAMPLE RÉSUMÉ UPDATE LETTER

JOHN DOE

224 Ernest Ave., / London, Ontario / W3E 4W3 / (555) 123-4567

FIRE TRAINING

Ice Rescue   **2007**
   Fire College   Toronto, ON

Confined Space   **2007**
   Fire College   Sarnia, ON

VOLUNTEER EXPERIENCE

Food Bank   **2007**
                                                                 London, ON

CHAPTER TWENTY-FIVE

VISITING THE FIRE HALL

You may ask yourself why it is important to visit the fire halls to which you wish to apply. If executed properly, visiting a fire hall can have a great impact on where you get hired as well as the amount of time it takes. You should try to visit every four to six months if possible. Your objective is simply to have the fire department(s) put a face to your name and to make them aware that you're interested and readily available.

Now that your objective is accomplished, the rest of your visit is filling in time with small simple questions. Focus is to ask questions like "Has it been busy today?", "How long have you worked here?", "What was the last fire you had while you were on duty?" and so on. I guarantee that asking simple questions will go much further and you will be remembered by the firefighter(s) when you come in for your next visit. Not to mention all the positive things that will be said about you to the other firefighters. In order for the rest of your visit to be successful don't get caught up bragging about yourself in depth. Limit the length of the visit to ensure the quality and impact of it. Shows you are a team

player. Remember the word "I" is not as important to the listener as when "you" is said.

Being self-involved or visiting too often you could become a bit of a nuisance. However, you do want them to continue to know that you're eager and available. To eliminate the possibility of being a nuisance you can show your appreciation by bringing in a tasty treat for the on-duty staff. This gesture will go a long way and you'll probably start a frenzy around the fire hall because everyone will be wondering where the treats came from and your name is going to be mentioned. This is exactly what you want. As far as time of day to visit a fire hall you should refrain from visiting during lunch or shift change hours. Most fire halls will be extremely receptive to you regardless of when or what time of day you happen to visit. If you are uncertain you could always call ahead and set-up an appointment to visit the fire hall.

During your visits you must have a positive attitude. Leave any negative energy at home. The last thing you want to have happen is to allow yourself to vent your personal frustrations on the nearest fire fighter and tell him/her about how hard it is to get hired or how you should have received an interview in the last recruitment for this fire department. If this conversation does arise say something positive, "It must not have been my turn to get hired. I know I'll get my chance; I just have to be patient. Do you have any ideas how I can improve my chances?"

In order to leave a good impression of your visit at the fire hall here are some key points to consider.

Appearance—Casual clothes are appropriate, such as jeans, shorts, T-shirts, etc. However, avoid wearing shirts that have suggestive or controversial statements on them. You should be clean shaven, well groomed and limited to little or no make-up. Also avoid heavy or strong scents. If you have piercing(s) or tattoos it's a good idea to conceal or remove them. However, not necessary but suggested.

Confidence—Be sure to have a firm handshake and make good eye contact. Ideally you should shake hands greeting and leaving the hall.

Manners—Make sure you demonstrate proper manners throughout your entire visit. Never chew gum or wear sunglasses when engaged in conversation or upon greeting someone. It's important they see your eyes to be convinced of your sincerity. A simple "thank you for your time" will suffice in closing your visit. On your way out of the hall be sure to thank everyone you see who talked to you during your visit.

Background—Before you decide to visit a fire hall try to gain some general knowledge about that department. Take the time to know the number of fire halls located in the city, types of calls they respond to and so on. This will demonstrate your interest level in that specific fire department.

Creating a relationship with a fire department is crucial. A common mistake candidates make is to restrict their visit to a fire hall only when that department is hiring. This makes it very obvious that you are only interested in the fire department because of the recruitment at the present time. In this case you may be better off not to visit at all and remain anonymous, hoping you will get called for an interview regardless.

I suggest that you start visiting fire halls in the areas you want to work in when they aren't hiring. This shows how obvious you are to want to work there and that you have some loyalty to that particular fire department. One of the hardest things about recruiting firefighters is cities do not only have to hire someone capable of doing the job, they also have to hire someone who is compatible with the firefighters at the fire department. Chemistry in the firehouse is extremely important and if you can prove that you're someone who is responsible, committed, loyal and has strong well developed interpersonal skills you would be seen by the fire department as someone who is a likely candidate to be recruited. Then all you have to do is back yourself up with an impressive résumé. If you do everything you can and keep doing everything you can, you should have your job offer.

It takes courage to grow up and turn out to be who you really are.
— Edward Estlin Cummings

! KEY POINTS TO REMEMBER

- If you are not receiving interviews then your résumé is not good enough. However, if you are receiving interviews and not getting job offers then you need to work on your interview skills.
- Don't forget interviews are on a topic you know all about—YOU.
- Cover letter should always accompany your résumé.
- Follow-up letters are a great idea after any interview, "the squeaky wheel gets the oil."

NOTES

_____

_____

_____

_____

_____

_____

_____

_____

_____

_____

_____

_____

_____

_____

NOTES

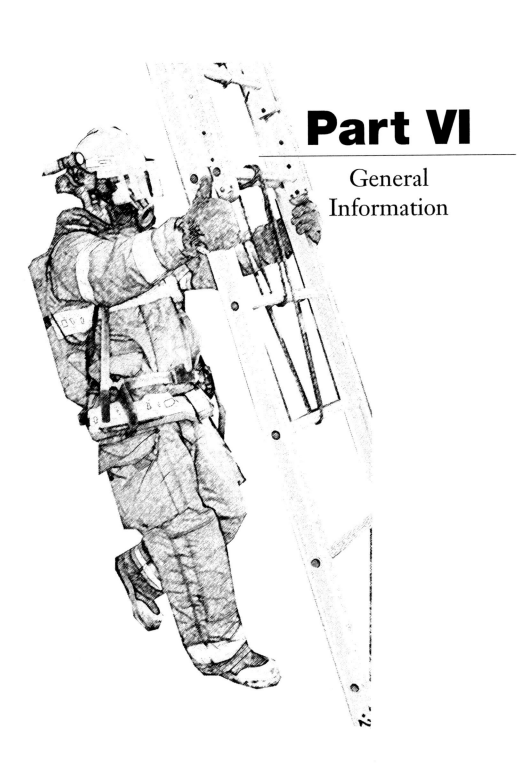

Part VI

General
Information

CHAPTER TWENTY-SIX

JOB SUCCESS

The day will come when all your hard work and effort will pay off. Your dreams of becoming a firefighter will come true. Every fiber of your being has been anticipating this moment since the day you decided to become a firefighter. Now the day has arrived. Chances are at this point you haven't given much thought about what you would do once you got hired or how you would act. The transition from someone who wanted to become a firefighter to someone who is a firefighter can be somewhat overwhelming at first. All of a sudden you're viewed by the public as a leader and someone to measure up to. You will leave your house knowing the chances are you may not come home the same person who left. On top of all this you have to integrate into a tightly woven family of firefighters at the fire house.

The key to job success in the fire department is simple; you must first and foremost respect your position as a probationary firefighter, fellow firefighters, and the "brass" meaning captains, lieutenants, officers, and chiefs. Even if you have a difference of opinion with someone you must respect their title. Secondly, remember, "We have two ears but only one mouth" You should be listening twice as much as you're speaking.

Always remember how hard you worked to get hired—never ever forget. Never allow yourself to become too comfortable or lazy. Always help others around the fire house, be courteous, work hard and make sure you keep your training at a level to influence saving your life or someone else's. Rewards come with hard work, and so does respect.

I can guarantee you that for your first weeks at the fire hall a few jokes will be played on you. Be careful not to let them get the best of you, keep your composure and don't let anyone think they are getting to you. Controlling anger shows maturity. Jokes are often ways meant to welcome you as one of the team. I would be more worried if the other firefighters didn't have fun with you.

WHAT'S EXPECTED OF YOU AS A PROBATIONARY FIREFIGHTER?

Roles and Responsibilities

Summary
- Take the job seriously
- Always be on time for work, classes, interviews, etc.
- Be honest
- Use common sense
- Be friendly and courteous to others
- Respect your co-workers
- Be responsible in public
- Be a good role model
- Keep up with your studies

Attitude
- Good attitude goes a long way in the fire service
- They want you to be someone they can train continually
- Keep the morale up at the hall
- Be someone that people like to be around
- Represent the fire department in all the good ways
- Be a leader in the community, volunteer
- Be someone who respects others

Firefighters don't have to be perfect. It's okay to make mistakes as long as you learn from them.

Probationary firefighter success tips:

- Promote honesty and integrity: these characteristics let others know they can depend on you.
- Demonstrate professionalism: you're a professional act like one.
- Check your personal gear (make sure it's all there), flashlight batteries, air pack, etc., at the start of every shift. You don't want to get a fire call and show up with an empty air cylinder.
- Be sure to check other equipment on the trucks before you start your shift.
- If a piece of equipment is damaged or broken tell your captain immediately.
- Be dependable, if you say you're going to do something—do it!
- Take the attitude "If I don't know or I'm not sure I'm going to ask instead of hurting myself or someone else".
- If you don't ask a question they assume that you already know; so be careful.
- If you have an exam you better be prepared, you don't want to do poorly on your exams. You are expected to ace your exams.
- Find a project: Look around at the equipment, station or crew quarters. Does something need attention? If so, repair it, clean it up, and leave it better than it was.
- When your duties are complete, help someone else complete theirs.
- Be a good listener, not only a story teller.
- Show up for work on time, make sure you're dressed appropriately and clean shaven.
- If your uniform is wrinkled, ripped or dirty take the proper measures to restore and maintain a professional appearance.
- Take care of equipment; be sure to put things in their proper place after you use them.
- Always appear busy or studying, you want to avoid the impression that you're goofing off or bored.
- Don't wait to be asked to do something, take the initiative.
- Watch what you say and who you say it to.
- Keep your mind active. Stay away from excess TV viewing.

Remember to have fun but take your job as a firefighter seriously, very seriously. Whether you're on or off duty you're representing every firefighter: past, present, and future. There is a bigger picture than you, just try to keep that in mind. When you're driving down the street and kids stop in their tracks just to get a glimpse of you and their eyes are bigger than silver dollars you will really start to understand the respected role of a firefighter in the community. It's important that kids continue to look at firefighters the way they do now—it's our job to preserve this.

NOTES

NOTES

Appendix

A – Schools

B – Websites

C – Magazines

D – Technical Books

APPENDIX A

FIRE FIGHTING SCHOOLS

Canada

AGI Shipboard Fire Service, Ltd.

Address: 561 Wain Road
           Sidney, British Columbia, Canada V8L5N8,
Tel: 250-655-3738
Fax: 250-655-3547
E-mail: shipfire@shaw.ca
Website: www.members.shaw.ca/shipfire/index.html
Program: Marine fire fighting for land-based fire departments, shipboard fire fighting for ship's crews, response training for fire watch personnel, command and control.

Algonquin College of Applied Arts and Technology

Address: 1385 Woodroffe Avenue
           Nepean, Ontario, Canada K2G1V8,
Tel: 613-727-4723
Program: Pre-Service Firefighter Education Program

Cambrian College of Applied Arts and Technology

Address: 1400 Barrydowne Road
           Sudbury, Ontario, Canada P3A 3V8
Tel: 705-566-8101
Email: info@cambrianc.on.ca
Webstie: www.cambrianc.on.ca

Program: Pre-Service Firefighter Education Program

College of the Rockies

Address: Box 8500, 2700 College Way
          Cranbrook, British Columbia, Canada V1C 5L7
Tel: 250-489-2751
Email: kolesar@cotr.bc.ca
Website: www.cotr.bc.ca
Program: Fire Training Certificate

Conestoga College of Applied Arts and Technology

Address: 299 Doon Valley Drive
          Kitchener, Ontario, Canada N2G 4M4
Tel: 519-748-5220
Email: webmaster@conestogac.on
Website: www.conestogac.on.ca
Program: Pre-Service Firefighter Education and Training

Dalhousie University, College and Continuing Education

Address: Dalhousie University, College and Continuing Education,
          Fire Management Certificate Program
          1535 Dresden Row, Suite 201,
          Halifax, Nova Scotia, Canada B3J 3T1
Tel: 902-494-3531
Email: gracetemani@dal.ca
Website: www.dal.ca
Program: Certificate in Fire Service Leadership; Certificate in fire Service Administration

Durham College

Address: 1610 Champlain Ave.
          Whitby, Ontario, Canada L1N 6A7
Tel: 905-721-3111
Email: don.murdoc@durhamc.on.ca
Program: Pre-Service Fire and Education Training Program

Emergency Services Academy Ltd.

Address: 2nd Floor, 161 Broadway Blvd.
          Sherwood Park, Alberta, Canada T8H 2A8
Tel: 780-416-8822
Email: esacanada@shawbiz.ca
Website: www.esacanada.com
Programs: EMR (Emergency First Responder—ACP Accredited),
          EMT (Emergency Medical technician—CMA Accredited and
          ACP Approved), Professional Firefighter (Fully Accredited including
          NFPA 472, 1001 Levels I &II Fire Fighter, and 1051, physical assessment)

Fleming College

Address: Sutherland Campus
          599 Brealey Drive, Peterborough, ON K9J 7B1
Tel: 705-749-5530
Fax: 705-749-5540
Website: www.fleming.on.ca
Program: Pre-service Firefighting Education and Training

Fire Etc. Training School

Address: 5704-47 Avenue
            Vermilion, Alberta, Canada T9X 1K4
Tel: 780-853-5800
Toll Free: 1-888-863-2387
E-mail: info@fire-etc.ca
Website: www.fire-etc.ca
Programs: Bachelor of Applied Business: Emergency Services, Emergency Medical
          Responder, Emergency Preparedness, EST — Emergency Services Technology
          Program, Extinguisher and Apparatus Maintenance, Fire Fighter Training
          Program, NFPA 472 — Professional Competence of Responders to
          Hazardous Materials Incidents, NFPA 1001 — Fire Fighter Professional
          Qualifications, NFPA 1002 — Fire Department Vehicle Driver/Operator
          Professional Qualifications, NFPA 1003 — Airport Fire Fighter Professional
          Qualifications, Canadian Aviation Regulations Standards (CARS) Training,
          NFPA 1021 — Fire Officer Professional Qualifications, NFPA 1031 —
          Professional Qualifications for Fire Inspector, NFPA 1033 — Professional
          Qualifications for Fire Investigator, NFPA 1035 — Professional Qualifications

for Public Fire and Life Safety Educator, NFPA 1041 — Fire Service Instructor Professional Qualifications, NFPA 1051 — Wildland Fire Fighter Professional Qualifications, NFPA 1081 — Industrial Fire Brigade Member Professional Qualifications, Safety Codes Officer in the Fire Discipline, Workplace Health and Safety

Georgian College of Applied Arts and Technology

Address: 1 Georgian Drive
          Barrie, Ontario, Canada L4M 3X9
Tel: 705-728-1968
Website: www.georgianc.on.ca
Program: Pre-Entry Fire Training (Community Service 1 year Program)

Holland College

Address: 140 Weymouth St.
          Charlottetown, Prince Edward Island, Canada C1A 4Z1
Tel: 902-629-4279
Toll Free: 1-800-446-5265
Email: info@holland.pe.ca
Website: www.hollandc.pe.ca
Programs: Para-medicine (CMA Accredited), NFPA 1001—Fire Fighter Level I & II

Humber College of Applied Arts and Technology

Address: PO Box 1900, 205 Humber College Boulevard
          Etobicoke, Ontario, Canada M9W 5L7
Tel: 416-675-6622
Email: ian.sam@humber.ca
Website: www.humberc.on.ca
Programs: Fire & Emergency Service Program

Justice Institute of British Columbia

Address: 715 McBride Blvd
          New Westminster, British Columbia, Canada V3L 5T4
Tel: 604-525-5657
Email: gmaddess@jibc.bc.ca

Website: www.jibc.bc.ca
Programs: Bachelor of Fire and Safety Studies, Career Fire Fighter Pre-employment
Certificate Program, Fire Service Leadership Diploma

Lambton College Industrial Fire School

Address: 1457 London Road
Sarnia, Ontario, Canada N7S 6K4
Tel: 1-800-791-7887
E-mail: info@lambton.on.ca
Website: www.lambton.on.ca
Programs: Pre-Service Firefighter Education and Training, 3 Year Fire Science
Technology Co-op Diploma Program (FST)

Manitoba Emergency Services College

Address: 1601 Van Horne Ave.. E.
Brandon, Manitoba, Canada R7A 7K2
Tel: 204-726-6855
Email: emerserv@gov.mb.ca
Website: www.firecomm.gov.mb.ca.
Programs: Fire Fighting Practices Program, Emergency Services Instructors Program,
Public Safety Program, Fire Prevention Program, Institutional Fire
Protection, Rescue Program, Driver/Operator Program, Fire Investigation
Program, Management Program, Building Standards, Emergency Medical
Program, Hazardous Materials Program

Nova Scotia Firefighting School

Address: 48 Powder Mill Rd.,
Waverley, Nova Scotia, Canada B2R 1E9
Email: nsfs@accesswave.ca
Website: www.nsfs.ns.ca
Programs: Pre-Fire Fighter Training, Fire Fighter I & II, Hazardous Materials
Response, Officer I & II

Offshore Safety & Survival Centre

Address: PO Box 4920
           St. John's, Newfoundland, Canada A1C 5R3
Tel: 709-834-2076
Toll free: 1-800-563-1344
E-mail: ossc@mi.mun.ca
Website: www.mi.mun.ca
Program: Firefighter Recruitment and Modular Training Program

Ontario Fire College

Address: 1495 Muskoka Road
           North Gravenhurst, Ontario, Canada P1P 1W5
Tel: 705-687-2294
Website: www.ofm.gov.on.ca
Program: Fire Prevention Officer and Company Officer Diploma

Portage College

Address: Box 417
           Lac La Biche, Alberta, Canada T0A 2C0
Tel: 780-623-5644
Website: www.portagecollege.ca
Programs: Wildland Firefighter, Wildland Operations, Wildland/Urban Interface
           Training

Ryerson University

Address: Ryerson University
           350 Victoria St., Toronto, Ontario, Canada M5B 2K3
Tel: 416-979-5057
Website: www.ryerson.ca
Programs: OFM-OAFC Partnership Program in Public Administration and Governance

Saskatoon Indian Institute of Technology

Address: #200-335 Packham Ave.
          Saskatoon, Saskatchewan, Canada S7N 4S1
Tel: 306-975-9636
Program: Emergency Services

Seneca College of Applied Arts and Technology

Address: 1750 Finch Avenue East
          Toronto, Ontario, Canada M2J 2X5
Tel: 416-491-5050
Website: www.senecac.on.ca
Programs: Pre-Service Firefighter Education and Training (1 Year), Fire Protection
          Technology (3 Years), Fire Protection Technician (2 Years)

St. Clair College of Applied Arts and Technology

Address: 2000 Talbot Road West
          Windsor, Ontario, Canada N9A 6S4
Tel: 519-972-2728
Email: www.stclaircollege.ca
Programs: Firefighter Pre-Entry Program, Paramedic Program

St. Lawrence College

Address: 2288 Parkedale Ave.,
          Brockville, Ontario, Canada K6V 5X3
Tel: 613-345-0660
Website: www.sl.on.ca
Program: Pre-Service Firefighter Program

UNITED STATES

Alabama Fire College

Address: 2501 Phoenix Drive, Tuscallosa, AL 35405
Tel: 205-391-3747
Website: www.alabamafirecollege.org

Delgado Community College

Address: 13200 Old Gentility Road, New Orleans, LA 70129
Tel: 504-483-4266
Fax: 504-483-4717
Toll free: 1-877-371-8206
Website: www.dcc.edu

Eastern Kentucky University Fire & Safety Engineering Technology

Address: 250 Stratton Building
          521 Lancaster Ave., Richmond, KY 40457-3131
Tel: 859-622-1053
Fax: 859-622-6548
E-mail: ipshopkins@acs.eku.edu
Website: www.fireandsafety.eku.edu

ESE Emergency Safety Environmental Training Associates, Inc.

Address: PO Box 427, Dalton, GA 30722
Tel: 705-342-5990
E-mail: info@prosafefire.com
Website: www.prosafefire.on.ca

Fire Science Academy (University of Nevada, Reno)

Address: 100 University Ave., Carlin, NV 89822-0877
Tel: 775-754-6003
Fax: 775-754-6575
Toll free: 1-800-233-0928
E-mail: fireacademy@unr.edu

Website: fireacademy.unr.edu

Kilgore College Fire Academy

Address: 1100 Broadway, Kilgore, Texas
Tel: 903/983-8662
E-mail: parrott@kilgore.cc.tx.us
Website: www.kilgore.edu/

Lake Superior Emergency Response Training

Address: 11501 Highway 23, Duluth, Minnesota State 55808
Tel: 1-800-432-2884
E-mail: arfft@computerpro.com
Website: www.isc.edu/
Programs: Fire Technology and Administration, Hazardous Materials and Safety

Louisiana State University Fire and Emergency Training Institute

Address: 6868 Nicholson Drive, Baton Rouge, LA 78020
Tel: 225-766-0600
Fax: 225-765-2416
Website: www.feti.isu.edu/

Maryland Fire & Rescue Institute

Address: University of Maryland, College Park, MD 20742
Tel: 1-800-256-3473
Fax: 301-220-0923
E-mail: adminsvc@mfri.org
Website: www.mfri.org

McMillan Offshore Training Center

Address: 148 Waterville Road, Belfast, ME 04195
Tel: 1-800-379-6678
E-mail: mmcmillan@mmcmillanoffshore.com
Website: www.mcmillanoffshore.com

Mississippi State Fire Academy

Address: Jackson, MS 39208
Tel: 601-932-2444
Fax: 601-932-2819
Email: fireacademy@msfa.state.ms.us
Website: www.doi.state.ms.us

Montana State University Extension Service; Fire Services Training School

Address: 2100 16th Ave.. S., Great Falls, MT 59406
Tel: 406-761-7885
Toll Free: 1-800-294-5272
Website: www.montana.edu

Montgomery County Fire Academy

Address: 1175 Conshohocken Road, Conshohocken, PA 19428
Tel: 610-278-3500
Fax: 610-278-3499
E-mail: RLINSINB@mail.montcopa.org
Website: www.montcopa.org/eoc/Fire_Academy/MCFAHome.htm

Oklahoma State University; Fire Protection and Safety Technology

Address: 303 Campus Fire Station, Stillwater, OK 74078-4082
Tel: 405-774-5721
Website: fpst.okstate.edu

Paul Hall Center

Address: PO Box 75, Piney Point, MD 50674-0075
Tel: 301-994-0010
Website: www.seafarers.org/phc/index.xml

Refinery Terminal Fire Company

Address: PO Box 4162, Corpus Christi, TX 78469
Tel: 361-882-6253

E-mail: info@rtfc.org
Website: www.rtfc.org

Resolve Fire & Hazard Response, Inc.

Address: PO Box 165485, Port Everglades, FL 33316
Tel: 954-463-9195
Fax: 954-356-5898
Toll free: 1-888-886-3473
Website: www.resolvefire.com

Salt Lake City ARFF Training Center

Address: PO Box 22107, Salt Lake City, UT 84122
Tel: 801-531-4521
Fax: 801-531-4514
E-mail: brian.pugh@ci.slc.ut.us
Website: www.ci.slc.ut.us

South Carolina State Fire Academy

Address: 141 Monticello Trail, Columbia, SC 29203
Tel: 803-896-9850
Fax: 803-896-9856
Website: www.llr.state.sc.us

Texas A&M University Emergency Services Training Institute

Address: 301 TarrowCollege Station, Texas 77845
Tel: 979-845-7641
Fax: 979-847-9304
E-mail: esti@teexmail.tamu.edu
Website: www.teex.com/esti

TSB Loss Control, Inc.

Address: 3940 Morton Bend Road, Rome, GA 30161
Tel: 706-291-1222
Email: tsblc@tsblosscontrol.com

Website: www.tsblosscontrol.com

Wyoming Fire Academy

Address: 2500 Academy Court, Riverton, WY 82520
Tel: 307-856-6776
Email: wfa@wyoming.com
Website: www.wyomingfire.org

APPENDIX B

FIRE FIGHTING WEBSITES

A
www.affm.mnr.gov.on.ca
www.allexpert.com
www.amazingfirephotos.com
www.americanfirefighter.com
www.applicanttesting.com
www.applicanttesting.com
www.atlanticfirefighter.ca

B
www.becomingafirefighter.com
www.bravest.com
www.brocku.ca/firefighter

C
www.cafc.ca
www.cafp.net
www.canwestfire.com
www.ccfmfc.ca
www.ciffc.ca
www.cfff.ca
www.city.toronto.on.ca/ems/image_files
    /btls.jpeg
www.coderouge.com
www.cwfis.cfs.nrcan.gc.ca

D
www.dartrescue.com

E
www.echelonresponse.com
www.emergency.com
www.emergencyservice.com

F
www.fcmr.forestry.ca
www.ffao.on.ca
www.fnfp.gc.ca
www.fiprecan.ca
www.firecanada.ca
www.firecareers.net
www.firedept.net
www.firedepartmentpatches.com
www.fireengineer.com
www.fire-ems.net
www.fire-etc.ca
www.firefighting.about.com
www.firefightingbrands.com
www.firefightingincanada.com
www.firefightinglinks.com
www.firefighterclosecalls.com
www.firefighters-memorial.com
www.firefighterprep.com

www.firefind.com
www.firefit.com
www.firehall.com
www.firehouse.com
www.firehouse651.com
www.firehydrant.org
www.firemuseumcanada.com
www.fire-resceutoys.net
www.firesafe.org
www.firesafetybow.com
www.firestore.com
www.firetactics.com
www.fsj.on.ca
www.fit-tech.org

G
www.geocities.com
www.glfc.forestry.ca
www.gov.bc.ca

I
www.ifsta.org
www.incendie.com

J
www.jems.com

L
www.let-r-graphics.com

M
www.magma.ca/~evb/forest.html
www.med-help.com

N
www.naemt.org
www.nfpa.org
www.nofc.cfs.nrcan.gc.ca
www.nsfirecism.ca

O
www.ontariofiretraining.com

P
www.permacharts.com
www.pfc.forestry.ca
www.planetclick.com

R
www.redcross.ca
www.rescue-net.com
www.rescuehouse.com
www.rescuetechniques.com
www.rocknrescue.com

S
www.stationone.ca
www.sja.ca
www.stokes-int.com
www.sparky.org

T
www.thefiredude.com
www.trainingcircle.ca
www.ttsao.com

W
www.wildfirenews.com
www.wildlandfire.com/jobs.htm
www.wfsi.org
www.workingfire.net
www.worldfiredepartment.com
www.911actionimages.com
www.911fallenheroes.org
www.9-11heroes.us

APPENDIX C

FIRE FIGHTING MAGAZINES

- Firefighting in Canada
- Canadian Firefighter and EMS Quarterly
- JEMS
- Fire Rescue
- Homeland First Response
- Firehouse
- Wildland Firefighter
- 9-1-1
- Fire Engineering

FIRE FIGHTING TECHNICAL BOOKS

IFSTA

FIREFIGHTER Item #: 36041
Essentials of Fire Fighting
Look for the Enhanced Version of Essentials—now includes full-color photographs! Note: The information in Essentials, 4th edition, is still the same. Addresses the 1997 edition and is correlated to the 2002 edition of NFPA 1001, Standard for Fire Fighter Professional Qualifications, Levels I and II, widely accepted as the standard of knowledge and skills measurement for all firefighters in North America and beyond. This IFSTA manual includes an appendix list of the job performance requirements from the NFPA standard and a cross reference of the NFPA requirements to the chapters of Essentials.

Essentials is the "bible" on basic firefighter skills and is the required training manual in countless local fire departments and state/provincial training agencies in every region of the country. It has an easy-to-read format with extensive use of photographs and colorful illustrations, plus 17 tables.

The use of skill sheets is new with this 4th edition of Essentials. Skill sheets describe the step-by-step procedures for many of the skills covered in the text. They are separated from the text to help make learning easier. Look for them at the end of many chapters. 4th Edition (1998) 716 pages.

Essentials of Fire Fighting Study Guide Item #: 36042
This study guide is a supplement to the fourth edition Essentials of Fire Fighting manual. The questions are designed to help students remember information and to make students think.

This guide has been separated into Firefighter I and Firefighter II sections to make the study process applicable to the various training entities that train to these levels. Most chapters contain material relevant to both Firefighter I and Firefighter II training.

The guide has a total of 2,037 questions. The Firefighter I section is broken down as follows: 614 multiple choice, 205 matching, 399 true/false, 258 identify, 7 case study, and 11 label. The Firefighter II section is broken down as follows: 184 multiple choice, 74 matching, 121 true/false, 79 identify, 13 list, and 11 label.

The answers, found in the back of the study guide, are listed by chapter and level. Each answer is referenced to the Essentials of Fire Fighting manual page on which the answer can be found. (1998) 404 pages.

Essentials of Fire Fighting Interactive CD-ROM (eBook) Item #: 36483
Essentials of Fire Fighting Interactive Book on CD-ROM.
(This is a single-user copy of Essentials of Fire Fighting on CD-ROM.)

Essentials of Fire Fighting Study Guide on CD-ROM Item #: 37115
These easy-to-use multimedia learning systems are modern, electronic versions of IFSTA's classic printed study guides. They allow the individual student to proceed through the questions in any order, find out immediately whether their answer is right or wrong and be referred to the source for the correct answer, if necessary. Most questions are in the multiple-choice format that is utilized by civil service and certification examinations. The program also tracks the results of each session as a record of progress. Each time the student goes through the program, the questions are rearranged randomly so each session has a different look. System Requirements: Windows(r) 95 and later, Windows(r) NT 4.0, Pentium Processor, 32 MB RAM, 17 MB Hard Disk Space, 16 bit High Color Display (minimum), CD-ROM Drive, Sound Card. In order for the program to operate on the user's computer, the CD must be inserted into the CD-ROM drive for verification.

Essentials of Fire Service First Aid Item #: 36569
Contains information that allows firefighters to learn minimum first aid requirements contained in NFPA 1001. The information was adapted from Fire Service First Responder which was produced by Brady. Special Edition.

Fireground Support Operations Item #: 36501
This manual replaces the IFSTA Fire Service Ventilation and Forcible Entry manuals and includes valuable information on fire service loss control as it applies to those two disciplines. Written for the advanced firefighter, the tools and techniques discussed in this manual are beyond those discussed in the Essentials of Fire Fighting, 4th edition, and they create an informational bridge to the Company Officer level.

In addition to discussions of fire behavior and size-up that are beyond those discussed in Essentials, this manual also discusses the most current ventilation methods and the most effective forcible entry techniques. The manual also uses numerous case histories as a means of stressing firefighter safety and survival on the fireground. These case histories provide examples of how firefighters have been injured or killed because they did not know or follow the principles and practices recommended in this manual. This is the manual on fireground "truck work," whether your department has a ladder truck or not. 1st Edition (2002) 324 pages.

Building Construction Related to the Fire Service Item #: 36295

The firefighter must understand building construction in order to understand the behavior of buildings under fire conditions. Meanwhile, building construction and materials are constantly changing. The firefighter must contend not only with the variety of buildings as they are currently being constructed, but also with a variety of buildings that were constructed with materials and methods that may now be obsolete. Unfortunately for fire suppression personnel, buildings may have been designed without regard for fire safety.

The 2nd edition provides the reader with basic information about how buildings are designed and constructed and how this relates to fire control and prevention. This new edition also includes case studies of the lessons learned from the behavior of buildings under fire conditions. This manual has value for fire inspectors, pre-incident planners, fireground commanders, and investigators as well as firefighters.

It includes numerous case studies, a glossary, and bibliography notes for each chapter and is fully illustrated with photographs and drawings.2nd Edition (1999) 212 pages.

Building Construction Related to the Fire Service Study Guide Item #: 36426

This study guide was designed to help the reader understand and remember the material presented in the second edition of Building Construction Related to the Fire Service. It contains more than 900 questions—165 matching, 223 true/false, 427 multiple choice, and 96 identify.

Class A Foam Item #: 36296

By Dominic Colletti

This is the most comprehensive book available on the subject of Class A foam, which is a chemical additive that when mixed with water forms a foam solution that is much more effective than plain water.

From the definition of Class A foam to how to use compressed air foam systems (CAFS), this book will increase your department's ability to stop a structure fire. According to the author, "The use of Class A foam and CAFS can enhance the fire suppression capability of water up to five times, thus dramatically improving the service it provides to its customers."

The text includes how to stop a structure fire in less time than using water, how the technologies improve firefighter safety, improvements in customer service, real world results using Class A foam, and more. 1st Edition (1998) 245 pages.

Fire Service Orientation and Terminology Item #: 36613

This new edition of Fire Service Orientation and Terminology acquaints new firefighters with a wide array of topics relating to the fire service. It describes the fire service as a career and explains the various roles of fire service personnel by illustrating the typical job and operation descriptions that should provide insight into the inner workings of the fire service. The manual also covers the traditions and history of the fire service. In addition to providing an overview of the fire service, Orientation and Terminology addresses such vital background work as fire prevention, firefighter safety, public fire and life safety education, and fire investigation. Also discussed are basic scientific terminology used in the fire service; basic building construction; and an overview of fire detection, alarm and suppression systems. It explains the relationship the fire service has with other organizations. It also covers fire department equipment and facilities, as well as fire department organization and management.

Orientation and Terminology has been revised to align with the National Fire Academy's Fire and Emergency Services Higher Education (FESHE) Initiative and meets the course objectives for Principles of Emergency Services (Introduction to the Fire Service). This manual also contains a fire service dictionary with thousands of terms used within the fire service. 4th Edition (2004) 433 pages.

Aircraft Rescue and Fire Fighting Item #: 36386

Aircraft Rescue and Fire Fighting addresses the requirements of NFPA 1003, Standard for Airport Fire Fighter Professional Qualifications, 2000 edition. The 4th edition of Aircraft Rescue and Fire Fighting provides basic information needed by firefighters to effectively perform the various tasks involved in aircraft rescue and fire fighting. Material covered includes qualifications for aircraft rescue and fire fighting (ARFF) personnel, aircraft and airport familiarization, firefighter safety, ARFF communications, rescue tools and equipment, ARFF apparatus and equipment, ARFF driver/operator, extinguishing agents, ARFF tactical operations, airport emergency plans, and hazards associated with aircraft cargo. 4th Edition (2001) 240 pages.

Fire Service Loss Control Item #: 36309

This manual covers traditional salvage and overhaul principles and skills and includes information to help fire service personnel understand a broader concept of loss control, which can be applied to all facets of fire service delivery emergency or non-emergency. Because efficiency, accountability, and professionalism are demanded by their community customers, loss control concepts are important to all fire service providers.

Fire Service Loss Control presents information in a chronological progression from pre-incident preparation to post-incident closure. The intent of the manual is not only to explain and demonstrate effective salvage and overhaul techniques, but also to reemphasize the importance of craftsmanship, pride in the profession, and compassion for those we serve in their time of need.

Fully illustrated. 1st Edition (1999) 117 pages.

Fire Hose Practices—8th Edition Item #: 36551

Fire Hose Practices, 8th edition, provides fire and emergency services responders with an in-depth resource on the most valuable tool for extinguishing fire: fire hose. Topics include the design and construction of various types of fire hose and couplings; proper procedures for cleaning, drying, repairing, storing, and service testing hose; and descriptions of the current generation of fire hose nozzles, appliances, and tools. The manual also contains a variety of methods for loading, unloading, advancing, and carrying fire hose during emergency incidents. 432 pages.

Aircraft Rescue and Fire Fighting Study Guide Item #: 36495

This study guide is designed to help the reader understand and remember the material presented in the IFSTA manual that it accompanies. The guide identifies important information and concepts from each chapter and provides questions to help the reader study and retain this information. In addition, the study guide serves as an excellent resource for individuals preparing for certification or promotional examinations.

Rapid Intervention Team Item #: 36482

By Gregory Jakubowski and Michael Morton

Rapid Intervention Teams was written by firefighters who have been implementing and teaching these concepts since 1995. The authors have worked in rural and urban environments and provide cost-effective solutions for your department.

Rapid Intervention Teams will:
*   Help guide you through the regulations and standards that apply to rapid intervention
*   Give you options for implementation and dispatch of rapid intervention units
*   Provide equipment lists for rapid intervention work, and show how work can be performed using existing equipment and training
*   Show a two-team concept that can provide the resources needed to handle the majority of rescue situations

1st Edition (2001) 183 pages.

Marine Fire Fighting Item #: 36352

The text addresses the shipboard fire fighting requirements of various maritime regulatory organizations such as U.S. Coast Guard, Canadian Coast Guard, and International Maritime Organization. It will aid the mariner, whether a new entrant about to join a vessel for the first time or an officer seeking higher certification, whether a young officer charged with inspecting fire equipment or conducting a training session or a senior officer conducting a drill while in command: All will find information within these pages to assist them in their tasks.

It provides a resource to the mariner for use when attending courses required by regulations in fire fighting and safety, for self-study, and when instructing others during training and drills. The content covers all aspects of fire prevention and suppression on board, whether a vessel is sailing deep sea, coastal, or inland waters or whether a mariner is in the merchant service, naval service, or coast guard. Vessel personnel must be their own fire department, police, and ambulance. This text will improve the basic and advanced knowledge and skill of shipboard crew members and also guide them when interfacing with land-based firefighters who respond to shipboard fires. 1st Edition (2000) 400 pages.

Marine Fire Fighting Study Guide Item #: 36420

This study guide is designed to help the reader understand and remember the material presented in the IFSTA manual that it accompanies. The guide identifies important information and concepts from each chapter and provides questions to help the reader study and retain this information. In addition, the study guide serves as an excellent resource for individuals preparing for certification or promotional examinations.

Marine Fire Fighting for Land-Based Firefighters Item #: 36476

The responsibility for fire suppression on vessels moored in or passing through an area belongs to the local jurisdiction. This text was written to provide training assistance to shoreside fire service personnel who respond to those fire emergencies. It covers subjects pertaining to the maritime environment, shipboard fire fighting strategies and tactics, and firefighter safety. It addresses the requirements of maritime regulatory organizations such as the U.S. Coast Guard, Transport Canada, and International Maritime Organization and also addresses NFPA 1405, Guide for Land-Based Fire Fighters Who Respond to Marine Vessel Fires.

Land-based firefighters need to understand their departments' policies and liabilities as they pertain to waterfront and marine incidents. Legal responsibilities and authorities aboard a vessel are different from those for a structure fire on land. This text describes the roles and responsibilities of responding agencies and the concept and implementation of a unified command structure. It also describes common situations encountered in vessel fire situations and characteristics of a shipboard fire incident plus vessel stability and ship/shore interface concerns. Vessel fires are significantly different from land-based structure fires. A vessel can move or capsize during fire fighting efforts, thus information on vessel structure, stability, and systems is essential to a successful fire fighting effort. The text contains pre-incident and tactical work sheets, stability calculation work sheets, technical data sheets, professional qualifications, and training standards ready for immediate use by fire service personnel. 1st Edition (2001) 440 pages.

Marine Fire Fighting for Land-Based Firefighters Study Guide Item #: 36496

This study guide is designed to help the reader understand and remember the material presented in the IFSTA manual that it accompanies. The guide identifies important information and concepts from each chapter and provides questions to help the reader study and retain this information. In addition, the study guide serves as an excellent resource for individuals preparing for certification or promotional examinations.

Respiratory Protection for Fire & Emergency Services Item #: 36502

This first edition of the IFSTA Respiratory Protection for Fire and Emergency Services manual takes the place of the Self Contained Breathing Apparatus (2nd Edition) manual that was specific to the fire service and limited to the use of SCBA. In developing this new manual, the IFSTA material review committee and the editor worked to include all types of respiratory hazards faced by personnel during emergency responses and the appropriate protection that must be worn. The committee recognized that changes in respiratory hazards, technology, regulations, and operational requirements justified a totally new approach to the subject and, therefore, adopted a new format for the manual. The manual is divided into two sections: the first dealing with administrative topics and the second with operational topics. While all of the information is needed for a comprehensive understanding of respiratory protection, the first section may be of greater value to personnel who must deal with selecting, purchasing, and maintaining respiratory protection in addition to facepiece fit testing. The operational section is directed toward the operational use, maintenance, and wearing of respiratory protection, along with emergency scene topics that interest the emergency responder.

Fire and emergency services organizations (public or private) must have an established respiratory protection program, including guidelines for use, maintenance, facepiece fit testing, and training. It is the intent of this manual to provide the necessary direction for such organizations in meeting the requirements for a respiratory protection program. 1st Edition (2002) 356 pages.

Wildland Fire Fighting for Structural Firefighters Item #: 36534

Written specifically for firefighters whose primary focus is fighting structure fires, but who are also responsible for protecting wildland (forest) and wildland/urban interface areas. This comprehensive text addresses all levels of NFPA 1051, Standard for Wildland Firefighter Professional Qualifications (2002 edition), and is consistent with the training standards of the National Wildfire Coordinating Group (NWCG). Replaces IFSTA's Wildland 3rd edition and earlier editions of the IFSTA Ground Cover Fire Fighting Practices.

Includes a laminated pocket card listing the 18 "Watch Out!" Situations, 10 Standard

Fire Orders, Common Denominators on Tragedy Fires, and LCES (Lookouts, Communications, Escape Routes, and Safety Zones). 4th Edition (2003) 528 pages.

Collapse of Burning Buildings: A Guide to Fireground Safety Item #: 35354

Based on 30 years of fire fighting experience in collapse situations, this book looks at analyzing causes of collapse and developing action plans for survival. Subjects include general collapse information, building construction, how buildings collapse, structural and collapse hazards, search and rescue, and safety precautions. 1st Edition (1988) Fire Engineering Books and Videos 287 pages.

Fireground Support Operations Study Guide Item #: 36570

This study guide is a supplement to the first edition of Fireground Support Operations manual. The questions are designed to help students remember information and make them think.

Foam Firefighting Operations 1—*The Essentials of Class A Foam* Item #: 36512

By Dominic Colletti and Larry Davis

This text is the first in a series of three volumes that focus on educating firefighters to various competency levels for the safe and effective use of Class A foam and compressed air foam systems. This book is highly valuable to fire organizations that either desire to improve their current use of foam technology or are considering the use of foam for the first time. 1st Edition (2002) 128 pages.

■ DRIVER/OPERTATOR

Pumping Apparatus Driver/Operator Handbook Item #: 36310

This manual is intended to educate driver/operators who are responsible for operating apparatus equipped with fire pumps. The information in this manual aids the driver/operator in meeting the job performance requirements in chapters 1, 2, 3, 6, and 8 of NFPA 1002, Standard for Fire Apparatus Driver/Operator Professional Qualifications, 1998 edition. It should be the goal of every fire department to train its driver/operators to meet all pertinent requirements contained in NFPA 1002. This is the training reference to use to reach that goal.

This manual combines all the important information previously contained in the IFSTA Fire Department Pumping Apparatus, Water Supplies for Fire Protection, and Fire Streams manuals needed by firefighters who seek to become driver/operators of fire apparatus equipped with a pump. Included are an overview of the qualities and skills needed by a driver/operator, safe driving techniques, types of pumping apparatus, and many more critical operations as shown in the chapter list.

Extensive appendix material includes blank daily, weekly, and monthly apparatus inspection forms; procedures for an appliance test method; friction loss calculation tables for U.S. and metric measurements; and a glossary of terms. Other tables provide information for troubleshooting during pumping operations. 1st Edition (1999) 471 pages.

Pumping Apparatus Driver *Operator Handbook Study Guide* Item #: 36320

This study guide is a supplement to the first edition Pumping Apparatus Driver/Operator Handbook. The questions are designed to help students remember information and to make students think. (2000) 184 pages.

Aerial Apparatus Driver/Operator Handbook Item #: 36393

While technology has brought a great amount of change to the fire service in recent years, one thing has not changed. The trucks still cannot extinguish fires by themselves. Competent, well-trained driver/operators, fire officers, and truck company members are essential for the optimum use of the aerial apparatus. The driver/operator must be thoroughly versed in the proper operation of the vehicle and the aerial device, the proper positioning of the vehicle, and the care and maintenance of the vehicle and its equipment. The fire officer and the truck company members must be prepared to perform any of the common truck company functions, including search and rescue, ventilation, salvage, forcible entry, exposure protection, and elevated master stream operations.

This first edition of the Aerial Apparatus Driver/Operator Handbook is designed to educate the driver/operators responsible for operating fire apparatus equipped with aerial devices. These include aerial ladders, aerial ladder platforms, articulating elevating platforms, telescoping elevating platforms, and water towers. It includes the information necessary to meet the job performance requirements of NFPA 1002, Standard for Fire Apparatus Driver/Operator Professional Qualifications, Chapters 1, 2, 4, and 5. It serves as a perfect accompaniment to IFSTA's Pumping Apparatus Driver/Operator Handbook for driver/operators who are expected to operate both types of apparatus. 1st Edition (2000) 284 pages.

Aerial Apparatus Driver/Operator Handbook Study Guide Item #: 36442

This study guide was developed to be used in conjunction with and as a supplement to the first edition of the IFSTA manual Aerial Apparatus Driver/Operator Handbook. The questions in this guide are designed to help you remember the information and to make you think. The study guide contains more than 800 questions in various formats: matching, true/false, multiple choice, identify, and put-in-order. (2000) 128 pages.

www.becomingafirefighter.com

Fire Department Pumping Apparatus Maintenance Item #: 36513

Produced by FPP

This text is the essential manual for both the emergency services technician and the operator. It elaborates and expands the knowledge and skills outlined in NFPA 1071, Standard for Emergency Vehicle Technician Professional Qualifications, by focusing on the inspection, servicing, maintenance, and repair of those systems that are unique to fire service pumping apparatus. It also covers the preventative maintenance required for fire department pumpers as outlined in NFPA 1915, Standard for Fire Apparatus Preventative Maintenance Program, by giving detailed explanations for how and why such a program needs to be implemented.

As an added bonus, the text also justifies NFPA 1901, Standard for Automotive Fire Apparatus, as it relates to maintenance. Included with the manual is a CD-ROM containing appendices on maintenance recommendations for major brands of fire pumps and inspection equipment, thus completing all of the information needed to maintain pumping apparatus to meet the NFPA standards. These appendices are not printed in the manual 1st Edition (2003) Fire Protection Publications 300 pages.

Fire Department Pumping Apparatus Maintenance Study Guide

Item #: 36514

Fire Department Pumping Apparatus Maintenance Study Guide

■ RESCUE

Principles of Vehicle Extrication Item #: 36382

In order to meet the challenges created by the technological advances in vehicle design and construction, this second edition of Principles of Vehicle Extrication is designed to serve as a reference in formal training courses on vehicle extrication and self-study by individual firefighters and other rescue personnel. This manual contains information on the newest types of air bags and other passenger-restraint systems as well as the latest tools and techniques used in vehicle extrication.

This manual exceeds the requirements of Section 4-4.1 of NFPA 1001, Standard on Fire Fighter Professional Qualifications (1997 Edition); addresses the requirements of Sections 3-1, 3-2, 3-3, 3-4, 3-5, and all of Chapter 6 of NFPA 1006, Standard for Rescue Technician Professional Qualifications (2000 Edition); and addresses the requirements of Chapters 2 and 6 of NFPA 1670, Standard on Operations and Training for Technical Rescue Incidents (1999 Edition). It is illustrated with hundreds of photos and illustrations. Considering that vehicle extrication incidents occur everywhere that land-based vehicles operate, there is a critical need for all rescue personnel to be fully aware of the challenges they face. If rescue personnel are to perform extrications safely and efficiently, they need the most up-to-date information and training available. 2nd Edition (2000) 208 pp.

Technical Rescue for Structural Collapse Item #: 36500

Structural collapse can occur in any jurisdiction at any time for a number of reasons. The fire department is usually the first to respond to these incidents. Therefore, rescue personnel must always be ready to locate and free victims from collapsed structures in the safest and most efficient way possible. This new manual, which addresses the structural collapse portion of NFPA 1006, Standard for Rescue Technician Professional Qualifications, and the 1999 edition of NFPA 1670, Standard on Operations and Training for Technical Rescue Incidents, is designed to go beyond the basic rescue skills detailed in the Fire Service Rescue manual to cover those needed by the rescue technician at these types of incidents.

Chapters include information on pre-incident planning and scene assessment, hazard reduction, safety, search techniques, cribbing, shoring, principles of lifting, mechanical advantage systems, lifting and moving techniques, rescue tool use for breaching and cutting, cutting techniques and safety, working with heavy equipment operators at the collapse site, moving victims and caring for them, securing and releasing the scene, and returning to post-incident readiness. 1st Edition (2003) 384 pages.

Fire Service Search & Rescue Item #: 36686

The purpose of the IFSTA Fire Service Search & Rescue, 7th Edition, manual is to provide emergency response agencies and their personnel with the information needed to meet the Operations-level requirements of NFPA 1670, Standard on Operations and Training for Technical Search and Rescue Incidents (2004). Since there are currently no Operations-level professional qualifications standards for technical rescuers, this manual identifies the requirements that agencies must meet, and the tools and techniques that their personnel must master if they are to provide Operations-level rescue service within their jurisdictions.

This manual focuses primarily on Operations-level situations to which firefighters and rescue squad members are most often called. However, to provide Operations-level fire/rescue personnel with the background information needed to effectively support and participate in complex technical search and rescue operations, some Technician-level information has been included. In addition, because firefighters constitute the primary intended audience for this manual, fireground search and rescue has also been included even though it is not addressed in NFPA 1670. 7th Edition (2005) 464 pages.

■ EMERGENCY MEDICAL SERVICES

Fire Service Emergency Care Item #: 36298

By Edward Dickinson, M.D. Published by Brady. Validated by IFSTA.

At last, a basic EMT textbook written specifically for the firefighter. IFSTA teams up with Brady to produce Fire Service Emergency Care. This unique text is the first basic EMT book ever to be validated through IFSTA. The author, Dr. Edward T. Dickinson, uses his personal background as a paramedic-firefighter and trauma physician to explore the difficulties of serving as first responders. The book is based on and complies with the guidelines established for the training of EMTs set by the Department of Transportation. The book, however, organizes and presents its contents with an awareness of the particular situations in which firefighters-EMTs work.

The handy company officer notes provide tactical, medical, and command warnings to the incident commander. Included are three appendices with additional information on ALS-assist skills, basic cardiac life support review, and National Registry EMT-B practical examinations. Extensively illustrated in full color. The comprehensive Fire Service Emergency Care is a necessity for every EMT-Firefighter. 1st Edition (1999) Brady/Fire Protection Publications 870 pages.

Fire Service Emergency Care Workbook Item #: 36299

Developed to be used in conjunction with and as a supplement to Fire Service Emergency Care. The chapters match those in the book. The questions are designed to help you remember information and make you think. The preface presents instructions on how to use the Workbook for maximum benefit. This Workbook also provides information about the National Registry EMT-B Practical Examinations. 1st Edition (1999) 316 pages.

Emergency Incident Rehabilitation Item #: 36575

By Edward T. Dickinson, MD, NREMT-P, FACEP
Michael A. Wieder, MS, CFPS

Emergency Incident Rehabilitation, 2nd Edition, is the most comprehensive and up-to-date presentation of rehab operations available. Authored by a member of the task group that developed the newly released NFPA 1584, Recommended Practices on Rehabilitation for Members Operating at Incident Scene Operations and Training Exercises, this edition has been thoroughly updated and expanded to meet these new standards. All of the basic functions that must be performed in a rehab operation are covered in detail and in a logical order, which allows them to be easily implemented by emergency operations of any size. In addition to meeting the NFPA 1500 and 1584 standards, the information contained in this book is also in agreement with the principles of the National Fire Service Incident Management System (IMS). 2nd Edition (2003) Brady/Fire Protection Publications 137 pages.

EMS Field Guide: Basic & Intermediate Version Item #: 36682

Produced by Informed

These handy field guides are the perfect on-site reference tool. The guides are waterproof, alcohol-fast, and tough to tear. Contents include everything from drug dosages to crime scene response. 5th Edition

EMS Field Guide: ALS Version Item #: 36684

Produced by Informed

These handy field guides are the perfect on-site reference tool. The guides are waterproof, alcohol-fast, and tough to tear. Contents include everything from drug dosages to crime scene response. 15th Edition

Emergency and Critical Care Pocket Guide Item #: 36683

Produced by Informed

These handy field guides are the perfect on-site reference tool. The guides are waterproof, alcohol-fast, and tough to tear. Contents include everything from drug dosages to crime scene response. 4th Edition

Fire & Rescue Guide Item #: 36688

Produced by Informed

The new 6th edition Fire & Rescue Field Guide includes new national ICS profiles, updated fire strategy & tactics, enhanced WMD and HazMat response, updated basic EMS section and improved operations center guidelines. This guide makes it easy for firefighters, command officers, and volunteers to coordinate fire attack, incident command, tactics, look up LELs, UELs, and friction loss. Now 7-color, this pocket-sized field reference is still only 3" x 5", has color-coded tabs, and is waterproof, alcohol-fast, durable & "Street Tough" 6th Edition

TO PLACE YOUR FIREFIGHTER BOOK ORDERS
PLEASE CONTACT:

ANNEX BOOKSTORE PH: 877-267-3473

VISIT ONLINE BOOKSTORE:
www.annexbookstore.com

PERSONAL CONTACTS

Contact name: _____

Occupation: _____

Address: _____

_____

_____

Phone number: _____

Fax number: _____

Email address: _____

Contact name: _____

Occupation: _____

Address: _____

_____

_____

Phone number: _____

Fax number: _____

Email address: _____

Contact name: _____

Occupation: _____

Address: _____

_____

_____

Phone number: _____

Fax number: _____

Email address: _____

Contact name: _____

Occupation: _____

Address: _____

_____

_____

Phone number: _____

Fax number: _____

Email address: _____

Contact name: _____

Occupation: _____

Address: _____

_____

_____

Phone number: _____

Fax number: _____

Email address: _____

Contact name: _____

Occupation: _____

Address: _____

_____

_____

Phone number: _____

Fax number: _____

Email address: _____

Contact name: _____

Occupation: _____

Address: _____

_____

_____

Phone number: _____

Fax number: _____

Email address: _____

Contact name: _____

Occupation: _____

Address: _____

_____

_____

Phone number: _____

Fax number: _____

Email address: _____

Contact name: _____

Occupation: _____

Address: _____

_____

_____

Phone number: _____

Fax number: _____

Email address: _____

Contact name: _____

Occupation: _____

Address: _____

_____

_____

Phone number: _____

Fax number: _____

Email address: _____

Contact name: _____

Occupation: _____

Address: _____

_____

_____

Phone number: _____

Fax number: _____

Email address: _____

ABOUT THE AUTHOR

I never had any ambitions to become a firefighter until my last few months of my apprenticeship. This was brought on after attending a firefighter recruitment seminar in my home town of London, Ontario. I came home that night with an overwhelming desire to be a firefighter. It took me two years to get hired as a firefighter for the St. Thomas Fire Department (start date: June 20th, 2003).

I started to write my manuscript for a firefighter recruit book days after being hired by the St. Thomas Fire Department. During my pursuit of becoming a firefighter I tried to research as much as I could about how to become a firefighter. But to my surprise there was no real governing authority on this subject and I found the market saturated with universal firefighting information. To prevent others from this situation I decided to become that authority. In 2004 I completed the book *Becoming A Firefighter: the Ultimate Recruit Guide For Canadian Firefighters*. I decided to print 500 copies of my book making me a self published author. I didn't stop there; I needed to provide more support to firefighter recruits and to accomplish this I created and host a website called www.becomingafirefighter.com.

My commitment to firefighter recruits has grown stronger as well as my presence as a resource for recruits over the last couple of years. I continuing to visit firefighting schools, write articles for international firefighting magazines and host my own firefighter recruit seminars. I remain the leading authority on becoming a firefighter.

Becoming a firefighter has definitely changed my life as well as my perspective on it. Devoting my time to help firefighter recruits achieve their dreams has become one of my favorite pass times.

Get the information to help make you the firefighter you know you can be!

To order your copy or a copy for a friend or loved one, fill out the Order Form below.

Don't forget to check out your local bookstore.

❑ YES, I want _____ copies of *The Complete Guide to Becoming A Firefighter* at $24.95 each, plus $6.15 shipping per book.

Enclosed is a cheque / money order in the amount of

$ _____ or bill my credit card:

VISA ❑   Credit card number: _____

Expiry date: _____

Signature: _____

PLEASE SHIP TO:

NAME: _____

ORGANIZATION: _____

ADDRESS: _____

_____

CITY: _____ POSTAL CODE: _____

PHONE: _____ EMAIL: _____

PLEASE MAKE CHEQUES/MONEY ORDERS PAYABLE AND MAIL TO:

KLP Services
6 Outer Dr.
London, Ontario
N6P 1C5

ALLOW 10 DAYS FOR DELIVERY.

Call your credit card order to: 519:652-8574
Email: info@becomingafirefighter.com